Designing for Longevity

Product longevity is one of the cornerstones in the transition towards a more sustainable society and a key driver for the circular economy model. This book provides designers, developers and creators with five distinctive expert strategies, detailed case studies, action guides and worksheets that support both beginning and advanced design practitioners in creating new product concepts with long-lasting strategic fits.

Designing for Longevity shows how expert design teams create original and long-lasting product concepts from the early development phase. It focuses on integrating business knowledge, market conditions, company capabilities, technical possibilities and user needs into product concepts to make better strategic decisions. It demonstrates how, for products to be durable, designers must create a long-lasting strategic fit for the customer, company and market. Key case studies of products such as Bang & Olufsen's A9, LEGO Ninjago and Friends and Coloplasts' Sensura Mio, among others, offer readers inspiration, guidance and real-world insights from design teams showing how the strategies can be applied in practice. Action guidelines and worksheets encourage broad, analytical problem-solving to identify and think through challenges at the early concept stage.

Beautifully designed and illustrated in full colour throughout, this book combines original research and the hands-on tools and strategies that design practitioners need to create useful, sustainable products.

Louise Møller Haase is an Associate Professor in the Department of Architecture, Design and Media Technology and Associate Dean of Education at Aalborg University, Denmark.

Linda Nhu Laursen is an Associate Professor in the Department of Architecture, Design and Media Technology and Head of the Design Lab at Aalborg University, Denmark.

DESIGNING FOR LONGEVITY

Expert Strategies for Creating
Long-Lasting Products

Louise Møller Haase & Linda Nhu Laursen

Routledge
Taylor & Francis Group

First published 2023
by Routledge
4 Park Square, Milton Park, Abingdon, Oxon OX14 4RN

and by Routledge
605 Third Avenue, New York, NY 10158

Routledge is an imprint of the Taylor & Francis Group, an informa business

British Library Cataloguing-in-Publication Data
A catalogue record for this book is available from the British Library

Library of Congress Cataloging-in-Publication Data
A catalog record has been requested for this book

ISBN: 978-1-032-28470-5 (hbk)
ISBN: 978-1-032-28466-8 (pbk)
ISBN: 978-1-003-29695-9 (ebk)

DOI: 10.4324/9781003296959

Typeset in Adobe Garamond Pro, Halyard Display, Benton Modern Display
This book has been prepared from camera-ready copy provided by Line Sand Knudsen

Access the Support Material: www.routledge.com/9781032284668

Acknowledgements

Writing this book has been an incredible journey to which many generous people offered their contributions.

First and foremost, enormous thanks to all the companies and designers who shared their expert knowledge and experiences – also the ones which are not directly mentioned in the book. Thank you all for your time, your openness and for sharing your truly exciting stories. Without you, there would be no book.

We would also like to extend an enormous thanks to Routledge for their professional and engaging collaboration when it comes to publish this book. Thank you for your expert guidance all the way from the first draft and throughout the final publication process.
Much gratitude also goes to the competent reviewers on the manuscript for fruitful perspectives, detailed feedback and engaging discussions related to the work.

Big thanks to the former "Designed to last" research group at the Industrial Design section at Aalborg University, and to colleagues in the research environment in general. Thank you for continuous discussions and conversations along the way.

Finally, we would like to acknowledge Line Sand Knudsen for her extensive work on this publication. She has made huge contributions to the final work of this book. Thank you for your thorough and committed help in the final writing process, for bringing the pieces into a whole, for your work on the graphics and layout of the entire book, and for your great contribution to the final publication process. Without your sharp attention to details and your solution-oriented attitude, this book would not have been what it is today.

CONTENTS

01 *Introduction*

02 *Circular economy and product longevity*

03   *Long-lasting strategic fit*

04   *Challenges when seeking a strategic fit*

05 *Expert strategies*

06 *Expert strategy 1*

07 *Expert strategy 2*

08 *Expert strategy 3*

09

Expert strategy 4

10

Expert strategy 5

11

Expert strategies and longevity challenges

12 *The road ahead*

13 *Action guides*

Design for product longevity

Strategic durability

A study of expert designers' practice

Why is this book relevant?

State-of-the-art research

The basic assumption and reasoning behind the research

Case selection

A short note on how we conducted the research

What's in store?

01

INTRODUCTION

Design for product longevity

Who would not want to create a product that is significant enough to last a lifetime? A product that customers love and admire? A product that proves to be relevant in the market for a decade or more and that makes a long-term strategic difference to the company that produces it?

Given the growing societal focus on achieving more sustainable production and consumption patterns, the creation of products that are designed to last is becoming increasingly relevant. In fact, we believe that in the coming years the ability to create long-lasting products will not only be a dream of many creators, but also a requisite skill for every designer, engineer and product developer.

Product longevity is a cornerstone of the transition towards a more sustainable society as well as a key driver of the circular economy model (Cooper 2010; Haffmans et al. 2018; Bakker et al. 2019). Today, 25% of all CO_2 emissions can be traced back to the creation of consumer goods (Ellen MacArthur Foundation 2021). However, by extending product lifetimes, product lifecycles can be slowed down, causing the number of products that need to be produced to decrease.

Design plays a central role in the creation of products with longer lifetimes. Research shows that up to 80% of a given product's environmental impact is determined during the design phase (Ellen MacArthur Foundation 2021). Therefore, it is crucial that the product lifetime is integrated as a key parameter in every aspect of the design process.

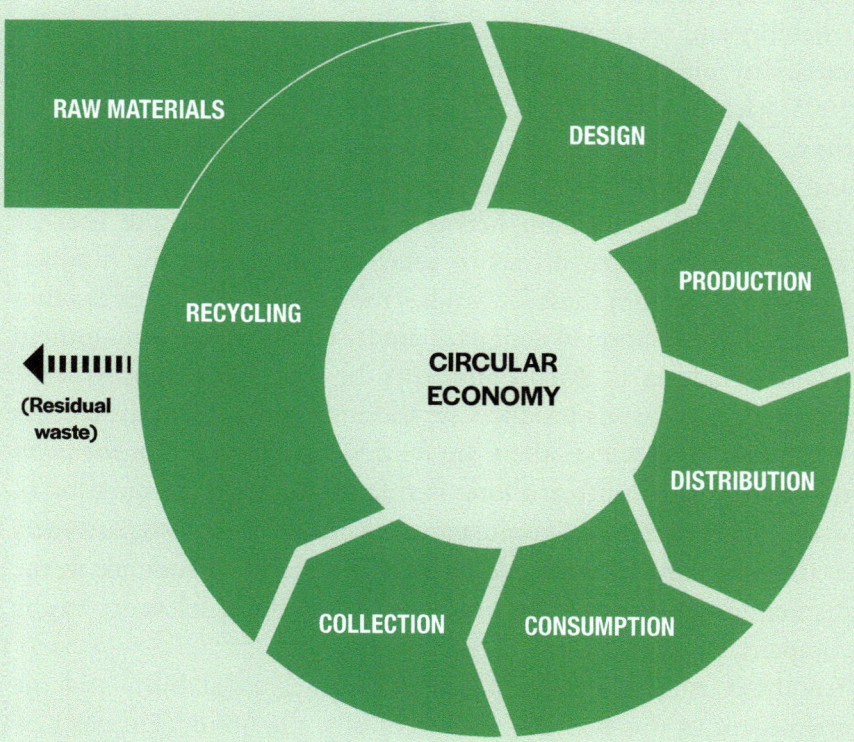

RAW MATERIALS

DESIGN

PRODUCTION

RECYCLING

CIRCULAR
ECONOMY

(Residual
waste)

DISTRIBUTION

COLLECTION

CONSUMPTION

*Figure 1.1: Circular economy model
(adapted from European Parliament 2015)*

Strategic durability

There are several design strategies available for achieving product longevity, including design for repair and maintenance, design for reliability and robustness, design for product attachment, design for variability and design for emotional durability (Van Nes and Cramer 2005; Moreno et al. 2016). The majority of these strategies focus on the technical, functional or material aspects of the product (Lofthouse and Prendeville 2017). Moreover, a couple of the strategies address the emotional connection between the product and the user, as well as the emotional drivers of using and discarding the product, although aside from these few studies the integration of the customer perspective in relation to design for product longevity remains limited (Haines-Gadd et al. 2018). Similarly, the competitive and strategic aspects of long-lasting products are seldomly addressed in the current strategies. For instance, there are no design strategies that focus on the strategic durability of a long-lasting product and the question of what it takes to render a product attractive and desirable to customers in the long run, generate a long-term competitive advantage in the market and ensure that the product has long-term relevance to the company.

Yet, these issues represent the core of strategic durability and the main focus of this book. More specifically, the book identifies new strategies that could serve to complement the current ones. While existing design strategies for product longevity provide valuable support during the later design phases, this book contributes with strategies that are particularly relevant to the strategic and conceptual phases of the design process, where many of the defining decisions concerning the product are made.

A study of expert designers' practice

This work was initiated due to our curiosity regarding how products with strategic durability are conceptualised and brought to life. During the research process, we realised that for a product to be 'strategically durable', it needs to have ***long-lasting strategic fits*** with the customer or user, the market and the company. In other words, a product must solve users' long-term problems and meet their needs, generate long-term competitive advantages in the market and be strategically important for the company. If the strategic fit is not long-lasting, the product will not be either. The reason for this is that if the product is not relevant for the user in the long term, it will be discarded prematurely. In fact, a study found that 42% of products are discarded while they are still 'fully functioning' (Van Nes 2003). Likewise, if a product is not competitive in the market or attractive to the company in the long term, it may be discontinued and, consequently, its value and lifetime may be drastically reduced. Over the last five years, we have engaged with more than twenty design teams from very different industries, who have all created some outstanding and strategically durable products. Although the expert design teams created very different types of products, their methods involved certain commonalities and shared strategies that seemed universal and highly important for all designers seeking to create products with strategic durability.

The expert design teams did not solely view their products in terms of the materials, features, functions and aesthetics involved. Instead, they looked deeply into the design principles associated with all products and understood how to adopt design principles from products that had proven to be strategically durable as a means of making their own product concepts strategically durable. Similarly, they were also highly conscious of the design principles that limit a product's strategic durability, which helped them to avoid making the same mistake in relation to future product concepts.

Hence, the principal aim of this research was to identify how expert design teams create long-lasting strategic fits and, consequently, ensure the strategic durability of the products they create. To achieve this, we investigated the design teams who have created the strategic concepts behind some truly remarkable and strategically durable products. Figures 1.2–6 present examples of the research cases that will be reviewed in this book.

Figures 1.2–6:
Examples of the research cases:
Vipp's V1 kitchen, LEGO's Friends series,
Bang and Olufsen's A9,
Coloplast's SenSura Mio and LEGO's
Ninjago series

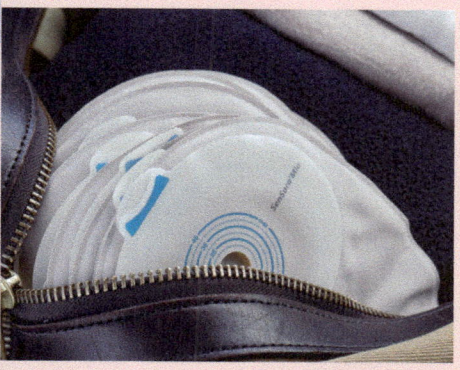

Case example
VIPP'S V1 KITCHEN
1.2. Credit: Vipp.com

Case example
LEGO NINJAGO
1.3 Credit: iStock.com/abalcazar

Case example
A9 - SPEAKER BY
BANG & OLUFSEN
1.5. Credit: Quang Tran

Case example
SENSURA MIO BY
COLOPLAST
1.4. Credit: Coloplast A/S

Case example
LEGO FRIENDS
1.6. Credit: iStock.com/Ekaterina79

Why is this book relevant?

This book has two main aims. First, we want to contribute to the research on product longevity and design for the circular economy by introducing a new research agenda that focuses on the strategic durability of products and sharing our key findings with expert designers. Second, because design is such an applied profession and we wish to have an impact on it, we have written a book that is also relevant and usable for practising design and development teams seeking to develop new lasting product concepts in a company setting.

Why is this book relevant to practitioners?

This book offers a number of benefits to designers and design teams who are developing new product concepts in a company setting and striving to design products with strategic durability. The task of creating a successful conceptual design represents a difficult challenge in itself. It is no trivial matter to combine knowledge regarding business opportunities, market conditions, company capabilities, technical possibilities and user/customer needs into a product concept. This is evident from prior research that has identified strategic conceptual design as the most complex and ambiguous aspect of the design process (Hey et al. 2007; Snowden and Boone 2007; Møller and Tollestrup 2013).

Product longevity further adds to the complexity of the conceptual process. However, it is crucial that considerations with regard to product longevity are included in this early phase of development, as decisions made during this part of the process are especially likely to impact the product's longevity.

With this book, we have done our best to accommodate the double complexity that characterises conceptual design and product longevity.

The book will disclose five strategies for achieving strategic durability that expert design teams can apply to create long-lasting strategic fits.

These expert strategies support the design of new product concepts that will be relevant to the user in the long run, generate long-term competitive advantages in the market, advance the long-term credibility of the company/brand, build on the company's strategic strengths and prove attractive for the company in the long run.

It must be acknowledged that the ability to create products with strategic durability requires both expertise and experience. This book will help readers to build up the required expertise. It presents five expert strategies for creating long-lasting strategic fits, as exemplified by five case studies from our dataset. Designers, developers and other practitioners can use the insights derived from these strategies and detailed case descriptions as guidelines for developing new product concept designs that strive to achieve strategic durability.

State-of-the-art research

Previous research concerning design for product longevity has mainly focused on the technical, functional and material aspects of products (e.g. Keoleian and Menery 1993; Bijen 2006; Vezzoli and Manzini 2008). For instance, there is extensive research on how to create products with long-lasting physical durability (e.g. Gupta and Gerchak 1995; Pangburn and Stavrulaki 2014; Tam et al. 2019) as well as on how to design products that are easy to maintain, repair, upgrade, etc. (e.g. Aziz et al. 2016; Khan et al. 2018). This focus is also reflected in many of the design strategies for product longevity listed in Table 1.7.

Design for durability **Design for reliability and robustness**	Strategies for creating products that are fit for the intended and foreseen use throughout their entire functional lifetime
Design for maintenance and repair	Strategies for creating products that allow for safe and quick cleaning, maintenance, replacement of worn-out components, etc.
Design for upgradability and adaptability **Design for variability**	Strategies for creating products that allow for replacing, adding or customising specific components
Design for standardisation	Strategies for creating products that can be connected to other devices, products or systems throughout their lifetime
Design for dis- and reassembly	Strategies for creating products that can be disassembled at the end of the product lifetime and then possibly reassembled into a new product

Table 1.7: Design strategies for product longevity
(adapted from Van Nes and Cramer 2005; Bakker et al. 2019)

Emotional durability and attachment

A product's lifetime is not only influenced by its ability to remain durable or functional, but also by the user's behaviour and perception of it (Bayus 1991; Grandberg 1997; Van Nes 2003; Cooper 2004; Oswald and Reller 2011). Products can become obsolete in the eyes of users and, consequently, be prematurely discarded, for example, due to their appearance deteriorating or their failure to keep up with trends or fashions. Products may also become obsolete once a new model is available on the market or when repair becomes too expensive in comparison to buying a new product (Cooper 2004; Burns 2010).

The challenge of product obsolescence has given rise to several strategies for addressing the use-oriented aspects of product longevity, including design for attachment, design for emotional durability and design for trust (Chapman 2005; Mugge et al. 2005; Schifferstein and Zwartkruis-Pelgrim 2008; Haines-Gadd et al. 2018).

The aim of all these strategies is to create an emotional bond between the user and the product, thereby making the user less likely to dispose of the product. This can be achieved by, for instance, creating a product that provides the pleasure of use, has an aesthetic appeal, stimulates memories or nostalgia, underlines self-expression, creates enjoyment or evokes sensory pleasure (Norman 2004; Cooper 2004; Mugge et al. 2005; Schifferstein and Zwartkruis-Pelgrim 2008).

Design strategies for slowing resource loops

Recently, the concept of product longevity has become central to the discussion concerning circular economy, a concept that aims to help companies and society shift away from the linear take-make-dispose pattern of production and consumption (Geng and Doberstein 2008) and towards a circular system in which the values of products, materials and resources are maintained for as long as possible and the amount of waste is minimised (European Commission 2020).

At the product design level, the move towards a circular economy is driven by two types of strategies: slowing and closing resource loops (Bocken et al. 2016). More specifically, strategies for slowing resource loops aim at extending the utilisation period of products, while strategies for closing resource loops aim at recycling materials and, therefore, closing the loop between discarded products and new production. In the context of circular economy, this means that product longevity is seen as a means of slowing resource loops.

Researchers have identified three key strategies for slowing resource loops (see Table 1.8) (Moreno et al. 2016; den Hollander et al. 2017; Bakker et al. 2019).

Design strategies for slowing resource loops		
Resisting obsolescence *Design strategies for long use*	**Postponing obsolescence** *Design strategies for extended use*	**Reversing obsolescence** *Design strategies for recovery*
Design for durability and reliability	Design for ease of maintenance and reuse	Design for recontextualisation
Design for emotional durability, attachment and trust	Design for upgradability and flexibility	Design for repair
	Design for standardisation and compatibility	Design for refurbishment
		Design for remanufacture
		Design for dis- and reassembly

Table 1.8: Design strategies for slowing resource loops and maintaining product integrity (adapted from Moreno et al. 2016; den Hollander et al. 2017; Bakker et al. 2019)

The first type of strategy is known as design for long use. The aim of this type of strategy is to design products that can resist obsolescence for as long as possible. The second type is design for extended use and its aim is to postpone obsolescence by, for example, designing products that are easy to maintain and upgrade. The third type of strategy is design for recovery. Here, the aim is to reverse obsolescence by, for instance, creating a design that is easy to repair or refurbish.[1]

Positioning: Design for strategic durability

With regard to current design strategies for slowing resource loops, the integration of strategic perspectives remains limited (Lofthouse and Prendeville 2017; Haines-Gadd et al. 2018). While researchers acknowledge that customer choices are not always rational and that they are influenced by a set of diverse and complex factors, including the alternative choices available on the market (Vezzoli and Manzini 2008), the current design strategies for creating long-lasting strategic products do not take into account the importance of strategic durability. This means that there is no guidance for practising designers concerning how to make a product attractive to customers in the long run, how to generate a long-term competitive advantage or how to ensure that the product will be relevant to the company in the long term.

The majority of initiatives intended to render a product able to resist, postpone or reverse obsolescence have not resulted in that product becoming more strategically durable. For instance, functionally durable products are often more expensive to both produce and purchase. While this may be acceptable in niche or monopolistic markets, if the customer is free to choose between different product alternatives, a long-lasting product must also provide some long-term benefits or superior value propositions if it is to remain attractive, competitive and relevant for the company (Van Loon et al. 2020; Alqahtani and Gupta 2017).

---

[1] *There are also a set of strategies that focus on dematerialising the product offering by, for example, creating a product service system or applying design strategies for swapping, renting and sharing (e.g. Morelli 2006; Vezzoli and Manzini 2008; Tukker 2015).*

The current design strategies for slowing resource loops provide valuable support for the later stages of the design process and for the detailed product design; however, they offer only limited insight into how product longevity should be handled during the strategic and conceptual phases of the design process, where many defining decisions regarding the product are made.

In this book, we focus on the strategic durability of long-lasting products and identify five design strategies for product longevity that can be applied during the early phases of the design and development process. In doing so, our aim is not to replace the current design strategies; rather, we seek to complement them.

We propose a new agenda within the field of circular product design that focuses on the strategic durability of products and, therefore, we identify design strategies for product longevity. In Table 1.9, we have positioned the research presented in this book in relation to other design strategies for slowing research loops.[2]

In summary, this book contributes to the literature by offering new knowledge on how expert design teams make products with strategic durability as well as concrete insights into their strategies for creating products with long-lasting strategic fits.

---

[2] *These strategies will not be further elaborated because this book focuses on the strategies for slowing resource loops while maintaining product integrity.*

Design strategies for slowing resource loops		
Resisting obsolescence *Design strategies for long use*	**Postponing obsolescence** *Design strategies for extended use*	**Reversing obsolescence** *Design strategies for recovery*
Design for durability and reliability	Design for ease of maintenance and reuse	Design for recontextualisation
Design for emotional durability, attachment and trust	Design for upgradability and flexibility	Design for repair
The research in this book: **Design for strategic durability**	Design for standardisation and compatibility	Design for refurbishment
		Design for remanufacture
		Design for dis- and reassembly

Table 1.9: Position of the present research in relation to other design strategies for slowing resource loops and maintaining product integrity (adapted from Moreno et al. 2016; den Hollander et al. 2017; Bakker et al. 2019)

The basic assumption and reasoning behind the research

The importance of closing and slowing resource loops is now widely recognised by politicians, researchers and society in general. Accordingly, different approaches to achieving a more sustainable future have been extensively discussed in terms of whether radical steps are required and how fast changes should be applied.

In the current literature, two main approaches for reducing environmental impacts have been identified, namely eco-efficiency and eco-effectiveness, which encompass two fundamentally different strategies with regard to circular economy (Penty 2020).

On the one hand, the eco-efficiency approaches aim to reduce environmental impacts through small, incremental steps that result in less waste, less energy and less material (e.g. Van Doorsselaer and Koopmans 2021). On the other hand, the eco-effectiveness approaches call for a more radical strategy that includes zero waste and toxicity, which can only be achieved if waste is used as a resource and everything is recycled (e.g. McDonough and Braungart 2002). The latter approaches require a fundamental shift in thinking and increased use of renewable forms of energy.

More recently, scholars have argued that circular design is an umbrella term covering several circular strategies and including a more systemic understanding of product design (Sinclair et al. 2018).

With this book, we aim to contribute to the creation of a design process whereby resource loops are both slowed and closed. Yet, we acknowledge that it can be very difficult, if not impossible, to identify 'perfect' or ideal cases of this process in practice. As a result, one could argue that a more conceptual or prescriptive research approach would be more suitable. Unfortunately, strategic durability cannot be studied in a vacuum such as in a lab or test setting. In practice, there are so many different stakeholders and dynamics that influence a product's strategic durability that any test situation, even if it involved the best setup, would limit this complexity. Hence, the study of strategic durability must be descriptive and based on real-life cases so that it can take the dynamics and complexities of the concept into account.

Case selection

The cases featured in this book are not intended to serve as perfect examples of circular products that will both slow and close resource loops. Rather, they have been included because they represent significant examples of products with strategic durability that have been 'resisting obsolescence' in markets characterised by fast phase change and, further, we believe that knowledge regarding how to achieve this could serve as an important supplement to current strategies for circular product design.

Moreover, our goal has always been to include as many cases as possible that do not focus on niche markets. Many studies concerning product longevity and the slowing of resource loops highlight niche market products with long lifetimes as perfect examples (e.g. Bocken and Short 2016). However, if the aim is to implement circular economy on a larger scale, we consider it relevant to study how strategic durability is created in larger and wider markets in which, for instance, price is also a key competitive parameter.

Table 1.10 summarises the cases featured in this book. It sets out the reasons for the specific case selections (i.e. the strategic durability indicators), the median product lifetime in each category and the expected product lifetime.

	Vipp V1 kitchen	B&O A9	LEGO Ninjago	Coloplast SenSura Mio	LEGO Friends
Strategic durability indicator	The Vipp V1 kitchen was introduced in 2011 and continues to be sold in its original form.	The A9 was introduced in 2012. The 4th generation of the product was marketed in 2019 and, today, first-generation A9 speakers are still sold at 50% of the original retail price.	LEGO Ninjago was introduced in 2011 and remains one of LEGO's main themes. LEGO Ninjago has become a collector's item. Unopened/sealed sets from 2012 have increased in value by an average of 330%. Used sets from 2012 are sold at 50–130% of the original retail price (BrickEconomy 2022).	The SenSura Mio was first marketed in 2014. Later, the main ideas behind the SenSura Mio were integrated into the general portfolio and remain significant to the company's product profile.	LEGO Friends was introduced in 2012 and remains one of LEGO's main themes. LEGO Friends has become a collector's item. Unopened/sealed sets from 2012 have increased in value by an average of 212%. Used sets are still sold at 20–40% of the 2012 retail price (BrickEconomy 2022).
Median lifespan of products in the category	Kitchens: 7 years (in 2005)	Speakers: 10.8 years (in 2005)	Toys: 4.7 years (in 2005)	Determined medically	Toys: 4.7 years (in 2005)
Expected lifespan of the products	25+ years	15+ years	20+ years	Determined medically	20+ years
			97% of all LEGO owners keep their old LEGO or distribute it to friends and family (LEGO 2022)		97% of all LEGO owners keep their old LEGO or distribute it to friends and family (LEGO 2022)

Table 1.10: Cases featured in this book

A short note on how we conducted the research

During the research process, we have had the pleasure of interacting with more than 20 expert design teams. Our focus was on learning about each design team's sensemaking in terms of the challenges they faced and the reasoning behind the design they created. Therefore, we conducted interviews during which we went through each design team's product creation process in detail. We discussed their early ideas and sketches, reviewed their presentations to management and looked through pictures and materials from various workshops.

These interactions allowed us to derive insights into the different approaches of the expert design teams, their findings, their mistakes and their successes. Most importantly, we obtained insights into their reflections, thinking and reasoning when it came to the key decisions regarding long-lasting products.

Interviews and discussions

To understand how the expert designers approached the different challenges they faced, as well as the strategies they used to turn those challenges into strategically durable products, part of the interview process focused on how the expert design teams framed every single aspect of the task, including how they framed the most important decisions during the product creation process. During the interviews, we asked the expert designers about how the different product aspirations were materialised into different solution principles. This involved questions such as the following: *What was the reason behind the product's visual expression? Why did it offer the experiences or interactions that it did? Why were certain technologies implemented when others were not? What were the reasons behind the material selection?* Next, we looked at the strategic aspirations of the product and asked questions such as the following: *How was it positioned on the market? How did it relate to the company?* We also encouraged the expert designers to share as many of their reflections, considerations and perspectives as possible. In fact, the questions we most commonly asked the expert design teams were as follows: *Why did you do this? How did you make sense of this? Why was this important? What was the reasoning behind this decision?*

Data analysis: The design DNA

During the data analysis, we focused on codifying the way in which the expert design team framed the product. We did so in accordance with framing theory (Schön 1983; Buchanan 1992; Schön and Rein 1994; Valkenburg 2000; Dorst and Cross 2001; Hey et al. 2007; Dorst 2015; Stompff et al. 2016). In this book, we will not detail the theoretical technicalities of framing theory or how we used it. However, if you are interested in learning more, please see our earlier work: Haase and Laursen (2019).

Slowly and after many hours of data analysis concerning each case, the 'framing' of each product emerged. We referred to this codified framing of the product as the 'design DNA'. The design DNA was important because it had guided all the design team's actions and decisions with respect to the product's development. After our analysis, the expert design teams reviewed and corrected the design DNA for each product in order to verify its validity. On the basis of the codified understanding of the designers' reasoning and sensemaking that the design DNA had provided, as well as the in-depth knowledge concerning the design process, we then started to identify patterns in the ways that the expert design teams had approached their design assignments. Within these patterns, we were able to identify both how the designers interpreted the strategic design challenges and the strategies they used to turn those challenges into strategically durable products.

What's in store?

Throughout this book, we have translated the knowledge derived from our work with the expert designers into maps, approaches and guidelines intended to clarify how the expert design teams make sense of challenges as well as the strategies they use to turn those challenges into strategically durable products. Furthermore, we have developed a number of suggestions for how to use the derived knowledge in a new design process.

Some of the material we collected during the research process is confidential because it reveals key company secrets. Fortunately, we are able to share five cases that exemplify the strategic challenges and strategies we identified. The five cases are Vipp's V1 kitchen, LEGO's Friends series, Bang and Olufsen's A9, Coloplast's SenSura Mio and LEGO's Ninjago series.

We realise that readers of this book are likely to range from beginners to experienced and advanced professionals. In light of this, we have dedicated equal focus to the strategies, case descriptions and detailed action guides, as they accommodate different learning styles and different levels of experience (Flyvbjerg 2006). Hence, we expect all readers to be able to access the type of knowledge they find most suitable. The following pages present an overview of the remaining chapters of this book.

Chapter 2
Circular economy and product longevity

In Chapter 2, we introduce the theoretical concepts of circular economy and product longevity. We review key aspects of the concepts, including product integrity, appropriate product lifetime, obsolescence, physical and emotional durability and circular business models. Chapter 2 forms the basis for understanding the rest of the book and, therefore, is a good place to start.

Chapter 3
Long-lasting strategic fit

In Chapter 3, we dig deeper into the notion of strategic durability and discuss how it can be created. We introduce the concept of **long-lasting strategic fit** and review how the creation of long-lasting strategic fits with the users, the market and the company leads to products with strategic durability.

Chapter 4
Challenges when seeking a strategic fit

In Chapter 4, we introduce **five different types of challenges** that designers typically face when they want to create products with strategic durability. Moreover, the chapter provides an overview of how to identify strategic challenges. This is handy for you, as a practitioner, when faced with the task of designing a new product or product line.

Chapter 5
Expert strategies

Chapter 5 provides an introduction to the **five expert strategies for creating products with strategic durability**, including the three key approaches: 1) renewing the strategic fit, 2) reframing the strategic fit and 3) framing the strategic fit.

Chapter 6
Expert strategy 1: Renewing core principles

In Chapter 6, we zoom in on **expert strategy 1**. The main focus of the 'renewing core principles' strategy is on identifying what caused an existing product to have a lasting product-user fit, product-market fit and product-company fit. The **Vipp V1 kitchen** is the case used to explain the strategy in practice.

Chapter 7

Expert strategy 2: Leveraging objections

In Chapter 7, we look closely at **expert strategy 2**. The main focus of the 'leveraging objections' strategy is on identifying the reasons why some products in the current portfolio have exhibited a weak strategic fit with the company and then using the knowledge derived from the unfitting products as a starting point for reframing the new product. The **LEGO Friends** case exemplifies this strategy.

Chapter 8

Expert strategy 3: Foreseeing future mismatches

In Chapter 8, we focus on **expert strategy 3**. The main focus of the 'foreseeing future mismatches' strategy is on identifying potential mismatches between what the user will find attractive in the future and what the company is currently offering and then using these mismatches as the starting point for reframing and renewing the strategic fit. The **B&O A9 speaker** is the case used in this chapter.

Chapter 9

Expert strategy 4: Extending product value

Chapter 9 is concerned with **expert strategy 4**. The main focus of the 'extending product value' strategy is on extending the value that the product provides to customers by, for instance, providing new services or covering new emotional or social dimensions that will allow the product to differentiate itself from other products and, therefore, consolidate its strong position in the market. **Coloplast's SenSura Mio** is the case reviewed in this chapter.

Chapter 10

Expert strategy 5: Searching for hooks

In Chapter 10, we concentrate on **expert strategy 5**. The main focus of the 'searching for hooks' strategy is on creating from scratch a lasting strategic fit with the user, the market and the company. **LEGO Ninjago** is the case of interest in this chapter.

Chapter 11
Application of the expert strategies

In this chapter, we discuss the **application of the expert strategies**. The five expert strategies for creating strategically durable products play an important role in the transformation from a linear economy to a circular economy.

Chapter 12
The road ahead

In Chapter 12, we discuss **the road ahead** for both practitioners and researchers. We draw connections between the different strategies and share some extra insights into the expertise required to create strategic durability.

Chapter 13
Action guides

This chapter presents the **action guides**, that is, the detailed descriptions of how to approach the practical design challenges associated with creating products with strategic durability. This chapter includes both guidelines and templates.

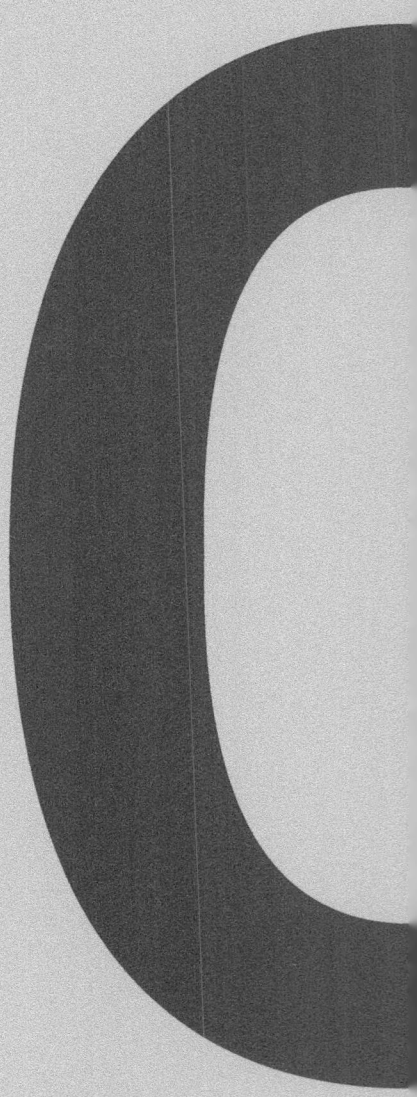

02

CIRCULAR
ECONOMY
AND PRODUCT
LONGEVITY

What is circular economy?

The importance of sustainable development has been recognised in academia for decades, while over the last few years such awareness and calls for action have also spread among politicians, businesses and people in general. In fact, sustainable development has never been as high on the global agenda as it is today. For most people, it is apparent that the current take-make-dispose pattern cannot continue. Indeed, this pattern has led to increasing pressure on both global resources and the climate due to human behaviour (Intergovernmental Panel on Climate Change [IPCC] 2014; World Business Council for Sustainable Development [WBCSD] 2021). Currently, humanity consumes renewable resources faster than they can be grown, leading to a scarcity of resources and the depletion of raw materials. This development is extremely critical and shattering for the climate, habitats and wildlife worldwide.

Circular economy (CE) is widely acknowledged as a promising approach to achieving a more sustainable future (Ellen MacArthur Foundation 2013; European Commission 2020). According to the Ellen MacArthur Foundation (2013), CE is 'an industrial system that is restorative or regenerative by intention and design. It replaces the "end-of-life" concept with restoration, shifts towards the use of renewable energy, eliminates the use of toxic chemicals, which impair reuse, and aims for the elimination of waste through the superior design of materials, products, systems, and, within this, business models' (8). Given this definition of CE, several questions may arise: What can we do in practice? How radical should the approaches we adopt to achieve change be?

Faber et al. (2009) found that the prior sustainability research can be divided into (i) studies that suggest we need to aim for an absolute or, perhaps, idealised form of sustainability and indicate radical measures to be necessary to achieve this and (ii) studies that recommend abandoning the aim of radical change and instead adopting a relative or pragmatic approach that begins with the present challenges and focuses on finding ways to address them.

In the design community, this division between the radical and relative approaches can also be seen. Since the 1980s, approaches such as eco-design, sustainable design, and design for the environment have been introduced in an attempt to encourage a more sustainable way of creating products (Moreno et al. 2016). However, these approaches have been criticised for not delivering the required change due to being built on a 'broken' system and, therefore, only delivering solutions that are 'less damaging' and 'less bad' (e.g. McDonough and Braungart 2002).

Yet, those who advocate for a pragmatic approach argue that small and incremental steps are required if new circular practices are to be implemented by customers, industry, public authorities, etc. For example, Van Doorsselaer and Koopmans (2021) argue that eco-design plays an essential role in the transition to a CE.

Despite these different approaches, CE is generally used as an umbrella term that includes all the circular strategies that envision the development of a more effective and efficient economic system through intentionally slowing and closing material loops (Bocken et al. 2016).

As shown in Figure 2.1, slowing and closing resource loops represent two fundamental strategies associated with CE (Bocken et al. 2016; Stahel 2016; Nußholz 2018). Walter Stahel is widely recognised as one of the pioneering developers of the slowed loop concept, which he introduced in 1982 in a paper entitled 'The product life factor'. Stahel (1982) suggested the extension of product lives as the starting point for a gradual transition towards a more sustainable society. Product life extension would slow down the flow of resources, thereby reducing the depletion of raw materials and the generation of waste. After a product's life has been extended for as long as possible, it should become part of a continuous system in which it becomes feedstock for new products. This can be achieved by looping the product back into the manufacturing process to allow for the reuse of its constituent components and the recycling of its materials (Stahel 2019). This process is known as the closed resource loop.

Thus, product life extension (slowing the resource loop) and recycling (closing the resource loop) are different yet complementary approaches to CE. While product life extension intervenes at the product level, recycling intervenes at the material level (Bakker et al. 2019). In the prior literature, design for recycling (the material level) can be seen as the predominant focus, whereas the product level and the concept of product longevity from a strategic perspective are largely overlooked. This book aims to foster a better understanding to this field.

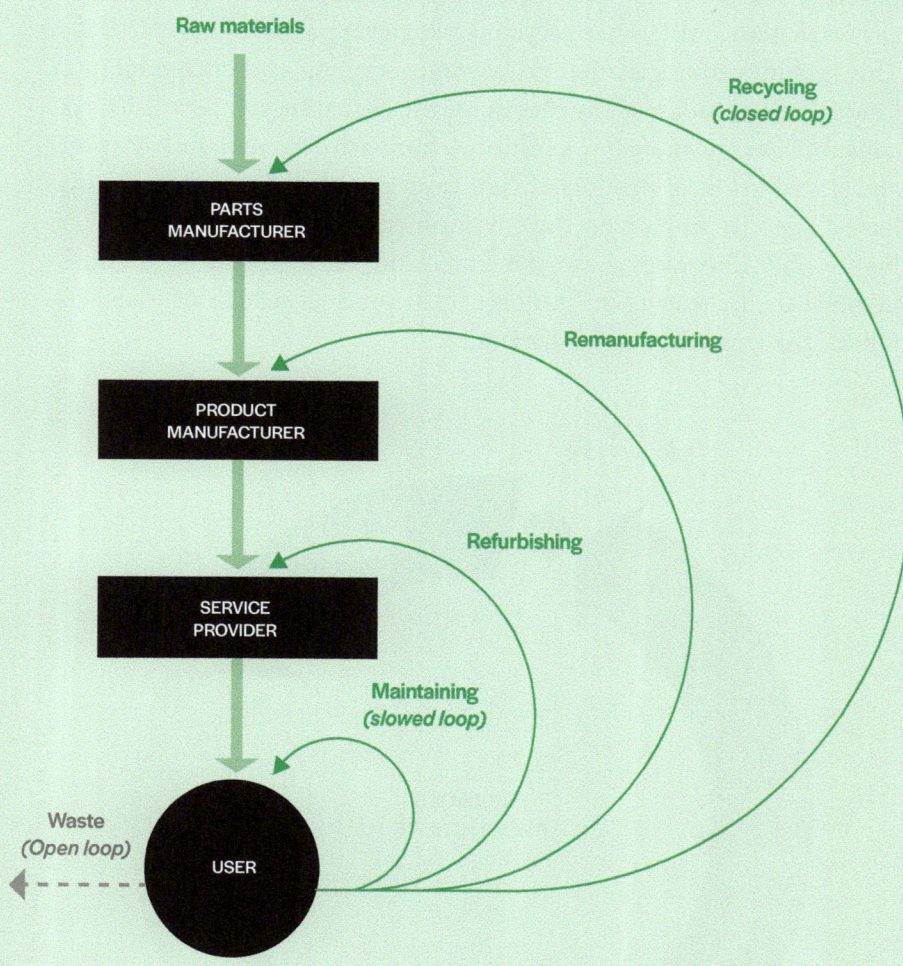

Figure 2.1: Slow and closed loops in a circular economy (adapted from Ellen MacArthur Foundation 2013; Bakker et al. 2019)

Why is product longevity important?

In terms of CE, several researchers have argued that the best option is always to reuse, rather than recycling whenever possible (Stahel 2019). This includes strategies that go beyond recycling and encourage the longer use of products. The reason for this is that recycling will always consume energy and cause pollution. Similarly, the remanufacturing and distribution of products made from recycled materials will also have a negative impact on the environment (Cooper 1994).

Figure 2.2 illustrates how the extension of a product's lifetime involves the increasing utilisation of that product, which will help to reduce the frequency of its disposal.

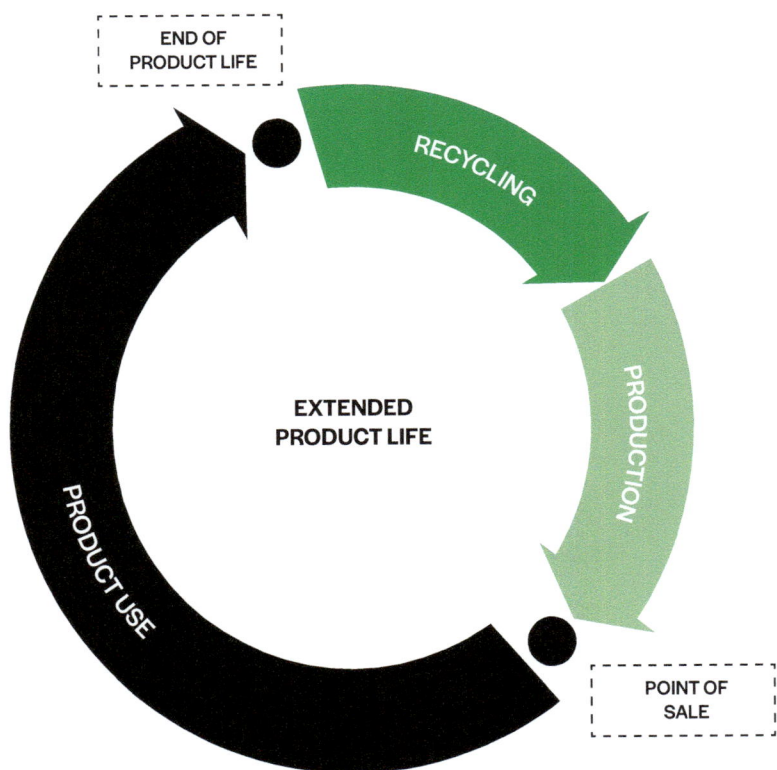

Figure 2.2: Extended product lifetime (adapted from Stahel 2019)

'Product longevity' is a key term used by product designers in the battle against the take-make-dispose pattern of production and consumption. It involves making better and fewer products that last longer. Accordingly, 'product longevity' is also a key term used throughout this book, although we first need to clarify its meaning. In the current literature, product longevity is generally considered to be a concept that covers four different yet interconnected issues: product lifecycle, product durability, product obsolescence and product lifetime (Jensen et al. 2021).

Product lifecycle refers to the entire process and series of stages of a given product, ranging from materials being processed, design, production, distribution, sales and use to disposal and recycling (in closed loop models). In the field of CE, a lifecycle assessment (LCA) is a widely used technique for calculating the overall environmental impact of a product, from its raw materials to the resultant waste. In Figure 2.2, a full circle depicts one product lifecycle.

Product durability most commonly refers to a product's physical properties, such as the quality of its materials, its ability to withstand 'wear and tear', and its capacity to perform its required function over a given period of time. We will examine this issue in more detail later in this chapter.

Product obsolescence refers to the user's emotional bond with a product. When a product has become obsolete, it is no longer considered relevant by the user (Burns 2010), which might result in the 'end of product life' (in Figure 2.1) even though the product is still able to perform its intended function. Later in this chapter, we will further elaborate on the issue of obsolescence and the various embedded emotional factors that are imperative to product longevity.

Product lifetime is the active lifetime of a product or the product's 'service life' (i.e. its total period in use). In Figure 2.1, the active product lifetime is depicted by the 'product use' arrow (the period from the 'point of sale' to the 'end of product life').

Given these definitions, this book defines product longevity in terms of products that *are **physically durable** and provide **long-term value** for users (emotional durability) while also representing a **viable business** and generating a **competitive advantage** for the company.*

These factors are all equally important and must be present at the same time to achieve product longevity.

Product longevity and product integrity

Another inherent aspect of product longevity is product integrity (den Hollander et al. 2017; Bakker et al. 2019). Bakker et al. (2019) argue that every change made to a product over time requires new energy and materials. Accordingly, a key aspect of product longevity involves maintaining the integrity of the product, which means that it should remain in its original state for as long as possible (Stahel 2010; den Hollander et al. 2017).

When a product is recycled (at the material level), its integrity is completely lost. If a product needs to be recycled, it is because it has been declared obsolete by its user for some reason. In other words, its use is no longer considered relevant.

Hence, product integrity first intends to resist obsolescence (at the product level) by keeping the product in use for as long as possible in its original state. Second, designers must ensure that the product can be recovered with the highest level of integrity (i.e. postponing or reversing its obsolescence) (den Hollander et al. 2017).

In Figure 1.5, we presented different strategies that can be applied to achieve product integrity at three different levels: 'resisting obsolescence' (highest level of integrity), 'postponing obsolescence' (medium level of integrity) and 'reversing obsolescence' (lowest level of integrity).

Previous studies have argued that product obsolescence can be prevented through the creation of products with high physical and emotional durability (Chapman 2009; den Hollander et al. 2017; Haines-Gadd et al. 2018). A long physical lifetime is clearly the key criterion for product longevity, but only if customers find the product valuable and desirable enough to keep and care for in the long run (emotional durability). In a similar vein, an emotionally durable product also needs to be physically durable if users are to keep and maintain it for a long time.

This book shows that strategic durability should be recognised as the third dimension of product longevity and product integrity. Indeed, a product needs to be competitive in the long term if it is to survive in a market characterised by an abundance of alternative solutions. It follows that if a product is not strategically durable, it will not be viable for the company and, accordingly, its physical and emotional durability will not be relevant either. All three types of durability must be present if a product is to have a long lifetime.

But what actually is a long product lifetime? How long is appropriate?

Appropriate product lifetime

The most appropriate lifetime from the CE perspective will ultimately depend on the type of product and its pattern of use. Hence, for a viable CE, the optimal product lifetime will vary, as not all products are suited to having a long lifetime. The creation of a long-lasting product often requires more effort and expense. Therefore, to achieve maximum effectiveness, it is imperative to identify the most appropriate lifetime and lifecycle scenario as early as possible during the design process (Penty 2020). As Cooper (1994) argues, from an environmental perspective, the goal is to ensure the optimum product life, not the maximum product life.

Products with the shortest lifetimes are considered to become waste after brief use, for example, hygiene products and packaging. Other product categories, such as information and communication technology (ICT) and other fast-paced technologies will inevitably be associated with shorter product lifetimes and so become obsolete more rapidly (Bakker et al. 2014). In such cases, product longevity may not be relevant, although material recycling becomes extremely important for a viable CE. Another example where a decreased lifetime is appropriate concerns products with high energy use, such as refrigerators. Prior studies have found that the frequent replacement of products that are always 'on' can represent an eco-effective strategy if they are replaced with a more energy efficient alternative (Kim et al. 2006; Van Nes and Cramer 2006).

At the other extreme, it could be argued that many antique items associated with outmoded social activities have been overbuilt and built to last for too long. There are also examples of 'over-engineered' products, that is, products with over-dimensioned utility or performance vis-à-vis their actual use, which might not be the optimal solution from the CE perspective.

Finally, there are products with which we associate product longevity. This often applies to products involving mature technology, such as cutlery and certain items of furniture, where the appropriate product lifetime is as long as possible. The same applies to the product cases included in this book, where the companies have been able to extend the product lifetime when compared with the median lifespan of the general product category (see Table 1.10).

Physical durability

When seeking to ensure the optimum lifetime of a product, a key consideration is the product's physical durability. In particular, product longevity is equivalent to high physical durability, meaning that the product's performance over time decreases at a slower rate than the performance of comparable products (den Hollander et al. 2017). Furthermore, physical durability represents a vital part of the user's experience of a product. For instance, a product with low physical durability may generate disappointment on the part of the user if it fails prematurely (Cooper 1994).

Therefore, designers must ensure the high quality of both materials and construction as well as the sturdiness and other physical properties of the product if it is to last. For example, a basic requirement involves preventing a product from wearing out due to the inherent limitations of the materials (Cooper 1994).

In the current literature on product longevity, design for durability is a widely discussed topic in terms of the physical properties of a product (e.g. Vezzoli and Manzini 2008). However, product longevity is not only a result of design and manufacturing decisions, as it is also influenced by customer behaviour and the emotional dimension associated with it (Cooper 2010). This implies that decisions regarding high physical durability are only relevant if the customer wants to keep and use the product in the long term (Chapman 2009).

Obsolescence and emotional durability

A product's lifetime is often defined in terms of its functionality; however, research has shown that many products are thrown away despite still being fully functional (e.g. Van Nes and Cramer 2005). Therefore, it has been suggested that a product's lifetime would be better defined in terms of obsolescence. This means that a product's lifetime is ultimately determined by the user, that is, it represents the time during which the product is perceived as valuable and significant by the user.

There are many different reasons why a user might decide to replace a product (Mugge et al. 2005; Van Nes and Cramer 2005). For instance, a user is likely to replace a product when its functional quality is inferior to that of newer models or when a supporting technology required to maintain the product is no longer available (Cooper 2004). Thus, obsolescence is the reason why a user might desire to replace a product. Accordingly, Burns (2010) described four different types of obsolescence that influence a product's lifetime:

Aesthetic obsolescence	Social obsolescence	Technical obsolescence	Economical obsolescence
Products may become outmoded due to: 1) worn appearance; and/or 2) new trends, styling or fashion.	Products may become irrelevant due to: 1) changes in behaviour or values in society; and/or 2) new standards (e.g. ISO), laws or regulations	Products may become outdated if the market introduces: 1) new models; 2) new technology; 3) new features; and/or 4) new levels of quality or performance.	Products may become obsolete if: 1) repair is too costly in comparison to buying a new one; and/or 2) there is a lack of availability of parts or repair shops.

Table 2.3: Four types of product obsolescence that influence a product's lifetime (adapted from Burns 2010)

aesthetic obsolescence, social obsolescence, technical obsolescence and economical obsolescence (see Table 2.3).

Obsolescence is highly influenced by manufacturers. In this regard, 'planned obsolescence' is a widely discussed term in relation to CE because it is a directly destructive strategy that leads to increased consumption and a throw-away society. Planned obsolescence is a business strategy that intentionally reduces the lifetime of a product and, therefore, encourages the user to 'own something a little newer, a little better, a little sooner than is necessary' (Adamson 2003, 4). This strategy is also referred to as the 'made to break' strategy because it can be achieved by intentionally making a product more fragile or difficult to repair than necessary (Rivera and Lallmahomed 2016; Ertz et al. 2019). Thus, it accepts the existence of waste, which stands in direct opposition to the concept of sustainable development.

Yet, companies could also potentially influence customers' decisions in a positive direction (from the CE perspective) by intentionally preventing obsolescence. To prevent obsolescence or product replacement, designers must consider a product's emotional durability.

Emotional durability is concerned with the user's experience of a given product. A product with high emotional durability fosters a long-term product-user relationship through the emotional connection between the user and the durable product (Haines-Gadd et al. 2018; Chapman 2021). In other words, to use the term that will be applied throughout this book, the 'strong fit' between the user and the product can serve to increase product longevity. If one experiences a strong bond with a product, one becomes emotionally attached to that product and, therefore, more likely to take care of it and keep it for longer (Mugge et al. 2005).

But how can designers influence the emotional bonds between products and users? Many 'design for X' strategies have been proposed to extend a product's lifetime, such as design for repair and design for refurbishment (e.g. Mugge et al. 2005; Van Nes and Cramer 2005; den Hollander et al. 2017; Bakker et al. 2019). While most of these strategies support the extension of the physical life of a product, only a few address the emotional life and the customer perspective.

In general, CE models have been criticised for not integrating the customer perspective sufficiently (Oghazi and Mostaghel 2018; Hankammer et al. 2019; Salvador et al. 2020).

The customer perspective represents an essential aspect of a company's strategy and business model, with the greatest challenge being to create a competitive value proposition that will render the long-lasting product both attractive and valuable to customers in the long run.

Circular business models

In practice, CE can be promoted and supported by the creation of business models (BMs) (Stahel 2010; Lewandowski 2016; Bakker et al. 2019). In the current literature, a business model conceptually describes 'the way business is done' (Magretta 2002) by qualifying what value is provided and to whom (value proposition), how that value is provided (value creation and delivery) and how the company makes money and captures other forms of value (value capture) (Bocken et al. 2014).

In the context of CE, a circular BM (when compared with the regular business models described in the business literature) incorporates a triple bottom line by 'generating competitive advantage through superior customer value while contributing positively to the company, the environment and society' (Bocken et al. 2015). Briefly put, a circular BM represents the rationale behind how organisations create, deliver and capture value in order to close or slow material loops (Antikainen and Valkokari 2016). In this regard, value is viewed in the broader context wherein collaboration between firms and other key stakeholders is necessary to deliver sustainability (e.g. Bocken et al. 2014; Manninen et al. 2017).

The central element that guides a BM is the value proposition (VP). This means that other elements of the BM (such as partners, activities and resources) are oriented towards the VP. From a customer perspective, the VP describes the benefits that the customer can expect from a given product or service (Osterwalder et al. 2014). From a company perspective, the VP reflects the company's core strategy for generating competitive advantages by offering value to specific target customers (Richardson 2008; Bocken et al. 2015).

Developing a VP for CE (particularly product longevity) involves a deep understanding of customers' long-term problems and needs. Moreover, it also involves determining the extent to which the company can satisfy those needs with a long-term solution that is superior to competitors' offerings. In other words, a sustainable VP is

Figure 2.4: Value proposition as the long-term competitive fit between customers' long-term problems and needs and the company's strategy, resources and competencies.

regarded as the fit between the company's offering and the customers' long-term needs (see Figure 2.4).

Although the customer perspective has been emphasised as a key parameter with regard to a viable BM, the environmental aspects are mainly considered within the CE literature. There are various examples of circular BMs that intend to extend product lifetimes but are rejected by customers when implemented (e.g. Kuo 2011; Haapala et al. 2008; Tukker 2015; Poppelaars et al. 2018; Zhou and Gupta 2019). In a comprehensive literature review, Tukker (2015) found various examples of product leasing or sharing models leading to less careful user behaviour and, further, identified that consumers often experience greater value from owning the products they use and appreciate having full control. Moreover, many consumers may only want to rent 'in-fashion' products (Besch 2005), which are short-living, meaning that this model may not represent a viable business for the manufacturer or contribute to sustainable development.

Even though these models exhibit great potential to extend a product's lifetime, they fail in practice because they are not attractive to consumers. This indicates the importance of circular BMs in not only supporting the extension of a product's active lifetime, but also in including a VP that is both desirable and attractive to customers in the long run.

Design strategies for different longevity business models

In the current literature, different business models have been proposed to increase product longevity in general and slow resource loops in particular. These models have been identified due to their focus on VPs (e.g. Bocken et al. 2014; Bakker et al. 2019).

There are two main models considered particularly relevant to product integrity and long use that also aim to resist obsolescence, which is the focus of this book. These models are the classic long-life and encourage sufficiency models.

Classic long-life

The VP in the classic long-life model is based on high-quality and long-lasting products as well as on delivering a high level of service over time (Bocken et al. 2016). Thus, it aims to maintain product integrity to the greatest extent possible by focusing on durable product design and supporting repair and maintenance from a long-term perspective. Accordingly, associated products are most often placed in the 'premium' price category in order to cover the long-term service and warranty costs.

A good example of the classic long-life model is Porsche. Indeed, most of the company's original cars are still on the road. The integrity of the old cars is maintained for many years, and even the spare parts have been restored in such a way as to maintain their originality. Many customers have a strong bond with their old cars and want to keep them even though newer cars might be more convenient, comfortable and, in some cases, cheaper to drive. In this regard, Porsche's products represent a good example of product longevity that considers both high physical and emotional durability. On the business side, Porsche has a specific program known as 'Porsche Classic' that takes care of the old cars and ensures they are in good condition and safe to drive while at the same time maintaining their originality. For instance, the company has a team dedicated to hunting down original spare parts from all over the world and making full restorations, including the original dip-coating for all Porsche cars.

It could be argued that other cars, such as the Fiat 500 or VW Beetle, are also classic models. However, while this is true, their long life solely concerns their image, as only few originals are still on the road, while the models have been continuously modified over time. At Porsche Classic, the spare parts represent a key aspect of the business model that supports a high service level in terms of repair and maintenance while still maintaining integrity for decades.

Encourage sufficiency

The sufficiency-driven BM is similar to the classic long-life model, although it follows an alternative direction when doing business (Bocken and Short 2016). The VP in the 'encourage sufficiency' model is based on solutions that actively seek to reduce the demand side, consumption and, therefore, production. On the business side, it actively encourages customers to consume less, for example, through a radical shift in promotion and sales, the provision of customer education and eschewing both discounting and overselling. Here, profitability is based on customer loyalty and increased market share due to product durability and longevity, which includes high-end and high-quality products within the 'premium pricing' category (Bocken and Short 2016).

A good example of the sufficiency-driven business model is the Vitsoe furniture company. The company is best known for its modular Dieter Rams shelving system from 1960, which remains part of its very limited product portfolio. Vitsoe promotes its products as an 'investment for life' and stresses that the furniture moves with you and adapts to changes in your life. Every single part of the system is made to last for as long as possible, and this includes lifelong service. The company is against fashionable products with limited lives, which means that you will never find a Vitsoe product on sale because the stock does not become obsolete. Vitsoe's vision of product longevity is very precisely described in its own terms: 'Living better with less that lasts longer' (Vitsoe 2022). The company's revenue model is based on a premium pricing model and its long-term and loyalty-based relationships with customers. Vitsoe clearly aims to change consumption patterns, which is key to the sufficiency-driven business model.

03

LONG-LASTING STRATEGIC FIT

Conceptual design

In the coming years, the design of products is likely to be highly influenced by the demand for more sustainable production and consumption patterns (Haffmans et al. 2018). Products will be expected to resist, postpone and reverse obsolescence as well as to fit into new types of circular BMs. The focus during the design process will be on slowing and closing resource loops. This will also be true of the conceptual design that is created during the early phase of the design process and in which the core synthesis of the functions, interactions, experiences, and strategies of the new product is performed (Møller and Tollestrup 2013; Andreasen et al. 2015).

During the conceptual design process, we foresee that the ability to create strategically durable products will be a prerequisite for many designers' practice and, further, that new types of strategic questions will drive this process. The following questions will all be of relevance: What will make the product attractive and desirable to the customer in the long run? What kinds of long-term problems will it solve for the user? How does it create competitive advantages that are relevant in the market in the long term? What will make the product strategically relevant for the company for many years to come?

Although strategic considerations are not uncommon during the conceptual design process, the nature of the strategic questions will probably change over the coming decades from what made sense in the linear economy to what will make sense in the CE. To explain this in more detail, we will present some of the prevailing theories concerning strategic and conceptual design in the linear economy and then show how the expert designers from our research alter them when they aim to create strategically durable products.

Strategic fit

In a linear economy, the creation of a new product concept is concerned with identifying new and emerging user needs, implementing new technologies, offering new and intriguing experiences, designing new kinds of interactions and creating new visual expressions, with the ultimate aim being to sell more products.

From a corporate perspective, 'product design' is a strategic tool used to approach the competitive challenge that the company is facing, for example, overcoming a competitive threat, maintaining market share, defending market position, maintaining technology leadership or entering a new market (Hooley et al. 2020). In general, the aim of a company within the linear system is to create a strategic fit between the needs of the customers, the conditions of the market and the company's resources, competencies and capabilities (Johnson, Scholes and Whittington 2008).

In a company, different strategies are applied to create or maintain the strategic fit. One such strategy for creating or maintaining a strategic fit is to create a new product or new product-line. This means that when designers are asked to design a new product or product concept, they are also asked (explicitly and implicitly) to contribute to overcoming one of the competitive challenges set out above. This process is illustrated in Figure 3.1.

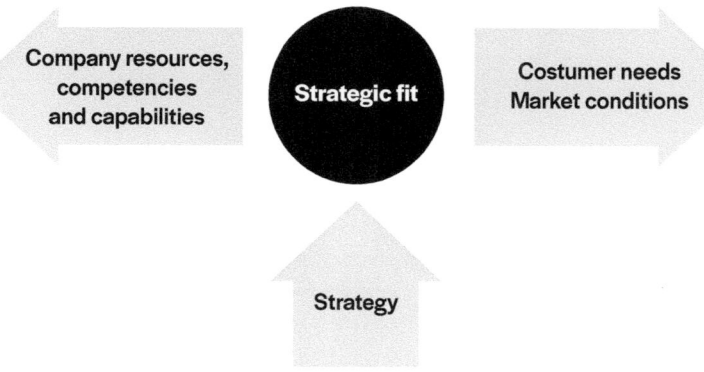

Figure 3.1: Strategic fit (adapted from Johnson et al. 2008)

Products as a means of creating a strategic fit

In most companies, the development of a new product is initiated due to the strategic fit (i.e. that created by the existing products) no longer being in place or sufficient. It could be that competitors are meeting the needs of customers better, the current product portfolio is outdated or the company has a new vision that needs to be set in motion (e.g. entering a new market).

The aim of the conceptual design process is, therefore, to create a new product that generates a strong product-user fit, a strong product-market fit and a strong product-company fit (see Figure 3.2).

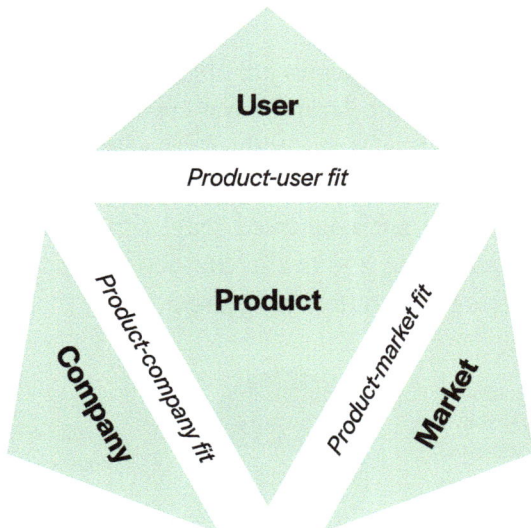

Figure 3.2: The conceptual design process aims to generate strong product-user, product-market and product-company fits

In more practical terms, the creation of a product with a strong strategic fit entails a design assignment that mandates the creation of a product that fulfils the needs, aspirations and wishes of users, creates a competitive advantage in the market and both advances the credibility and enhances the strategic strengths of the company. In the following sections, we will discuss this further.

The product must fulfil the needs, aspirations and wishes of the user

To create a strong product-user fit, products must fulfil user needs (Laurel 2003; Merholz et al. 2008) and be meaningful to users (Krippendorff 2006). Sanders and Stappers (2014) further argue that it is important to look beyond users' explicit needs. Rather than asking users what they currently need or may need in the future through surveys and interviews, they suggest that it is important to develop deep insights into users' behaviour, aspirations, dreams and wishes. Ulwick (2005) argues that it is vital to understand the functional, personal and social roles that the product plays for customers. Any car can transport a person from A to B, but how the person feels while driving the car (personal role) and how other people think about the person driving it (social role) can be quite different depending on the specific brand of car. Hence, to generate a strong product-user fit, designers must create products that meet users' needs, wishes, dreams and expectations and also support users' behaviour.

The product must create a competitive advantage in the market

To create a product-market fit, a product must provide a set of competitive advantages that will ensure that customers choose it over competitors' products (Hooley et al. 2020). Porter (1980) argues that competitive advantages can only be created through either cost leadership, differentiation or a combination of the two. Cost leadership is typically used in commodity markets, although here it is created by, for example, efficient production processes or superior production technology that lowers the production costs. Differentiation can be achieved by creating something that is considered unique on the market, for instance, new product experiences or visual expressions (style), new ways of interacting, new product or service features and differentiation within a brand or distribution channel.

Hence, the creation of a product-market fit depends on the design team's ability to create a product with unique and differentiating features that generate competitive advantages and are highly difficult for competitors to copy or imitate.

The product must advance the credibility of the company

To create a strong product-market fit, it is also important that the product advances the credibility of the company and the brand. The company's competitive position influences the types of products it is considered credible for the company to market. For instance, if a company lacks a high-quality brand, there is a limit to what customers are willing to pay, even if it delivers a high-quality product. Likewise, if a brand is known for its safety, it will only take one unsafe product to ruin both the perception of safety and the competitive position of the brand.

According to Kotler (2003), 'Positioning is the act of designing the company's offering and image so that they occupy meaningful and distinct competitive positions in the target customers' minds.' Products are just one of many factors that influence or establish a company's competitive positioning in the minds of customers. In fact, customers compare different companies, brands and products to each other, thereby developing an understanding of how the companies or brands are positioned relative to each other, for instance, when it comes to value for money, quality, meaning or image.

Thus, generating a product-market fit depends on the design team's ability to create a product that advances the credibility of the company and, therefore, underlines or strengthens its competitive position in the market.

The product must enhance the strategic strengths of the company

It is highly important to create a strong product-company fit too (Holston 2011; Buijs 2012; Curedale 2013). It is vital that new products are built on the company's core competencies and strategic strengths. Most companies have core strengths or distinctive competencies that are superior to those of their competitors. These can entail specific technologies, production methods and distribution channels or specific knowledge or competencies regarding materials, markets, etc. Therefore, it is highly advantageous to build on these strengths and competencies, if possible.

Moreover, as it is both costly and risky to change technologies, production processes or distribution channels, such changes should be carefully considered before they are made a prerequisite for a new product or product line. As Buijs (2012) argues, 'Innovation is risky. Therefore, it is better to build on the strategic strengths of the company, rather than on resources that every company has. It does not look very clever to build a new strategy on the company's weaknesses' (60).

Hence, generating a product-company fit depends on the design team's ability to create a product that takes advantage of the distinctive competencies that render the company superior to its competitors.

The product must align with the company's values, purpose and culture

To create a strong product-company fit, Collins and Porras (1996) suggest that it is important to take into account the company's overall vision, values and culture, which are all highly influential when it comes to what is identified as attractive for the company. In addition, according to Collins and Porras (1996), 'Companies, that enjoy enduring success have core values and core purpose that remain fixed while their business strategies and practices endlessly adapt to a changing world' (65).

A company's purpose is its reason for existing. It also represents the idealistic reason and motivation for people to work for the company.

A company's purpose should not be confused with its current vision or goals. A purpose is long-lasting and always truthful. 'It is like a guiding star on the horizon—forever pursued but never reached' (Collins and Porras 1996, 86).

A company's core values are a set of timeless guiding principles that it navigates by. The core values express what is important to the company regardless of any changes in the market or competitive environment. The core values require no justification because they are essential tenets of the company. Thus, generating a product-company fit depends on the design team's ability to create a product that aligns with the company's overall purpose and values.

In summary, when design teams are assigned to design a new product, they are also (explicitly or implicitly) asked to create a strong strategic fit with the user, the market and the company. This means that the new product must:

- fulfil the needs, aspirations and wishes of the user;
- create a competitive advantage in the market;
- advance the credibility of the company in the market;
- enhance the strategic strengths (e.g. resources, competencies, capabilities, technologies, know-how, etc.) of the company; and
- integrate the company's purpose and values in an efficient way.

Strategic durability and long-lasting strategic fit

In a linear economy, the main emphasis during the conceptual process is on creating a strong strategic fit. Some would even argue that creating a strong strategic fit may form part of a planned obsolescence strategy on the part of certain companies, where new products encourage customers to exchange their current products a little sooner and faster than necessary.

This also means that the current theoretical models for creating a strategic fit are not sufficient. Indeed, when it comes to creating long-lasting products with the aim of resisting obsolescence, the current models fall short. Although the design, marketing, entrepreneurship and corporate strategy literature addresses various ways to create different types of strategic fits, no strategies for creating strategic fits for products intended to have long lifetimes can be found in the current literature.

Fortunately, through our research on expert design teams, we have identified how such teams manage this in practice. We have seen how they are very conscious of the need to create products with strategic durability that can resist obsolescence as well as of how their strategies and approaches differentiate their process from the regular conceptual design process.

The main thing that differentiates the processes followed by the expert design teams from the processes discussed in the current linear literature is the fact that they are not only highly conscious of creating strong product-user, product-market and product-company fits, as they also focus on creating *long-lasting* strategic fits.

From our research, it is evident that, for instance, the expert design teams do not solely attempt to identify and fulfil any types of user needs, wishes and aspirations. Rather, they search for *long-lasting* problems to solve as well as for *long-term* needs and aspirations to fulfil, and they pay attention to *long-term* personal and social needs in particular. As one of the expert designers explained, 'we need to find out what really resonates with users'.

Likewise, it is evident that the expert designers seek to create long-term competitive advantages, not just competitive advantages of any duration. That is, they want to create advantages that are not based on fast-moving trends or other kinds of temporalities, but on unique features that are hard to imitate and relevant in the long term. To this end, the expert design teams are very conscious of how the new product would advance the long-term credibility of the company.

Finally, our research also shows that the expert design teams not only build on the company's core competencies and strategic strengths, but also aim to enhance them. They are highly engaged in terms of aligning the product concept with the company's values, purpose and culture, thereby rendering the new product attractive to the company for many years to come.

Based on the findings of our research concerning the expert design teams, we propose the following:

Creating products with strategic durability entails creating products with long-lasting strategic fits. That is, products that meet long-term user needs, create long-term competitive advantages in the market, advance the long-term credibility of the company, enhance the strategic strengths of the company and align with the company's values, purpose and culture.

This proposition is illustrated in the long-lasting strategic fit map shown in Figure 3.3.

LONG-LASTING STRATEGIC FIT MAP

Fulfill long-term user needs, aspirations and wishes
Products with long-lasting strategic fits solve long-term problems for users and fulfil the needs, wishes and aspirations of users in a way that is relevant both now and in the long term.

Enhance the strategic strengths of the company
Products with long-lasting strategic fits enhance the company's core competencies and strategic strengths. The products take advantage of the distinctive competencies that render the company superior to competitors but also advance them.

Create a long-term competitive advantage in the market
Products with long-lasting strategic fits create long-term competitive advantages for the company, typically by offering unique and differentiating features that are long-lasting and highly difficult for competitors to copy or imitate.

Align with the company's values, purpose and culture
Products with long-lasting strategic fits align with the company's purpose and values in order to remain attractive to the company for as long as possible. At their best, such products clarify, materialise and bring new life to the company's values, purpose and culture.

Advance the long-term credibility of the company
Products with long-lasting strategic fits advance the long-term credibility of the company and exert a long-term positive impact of customers' perception of the company, thereby underlining and strengthening its competitive position in the market.

Figure 3.3: Long-lasting strategic fit map

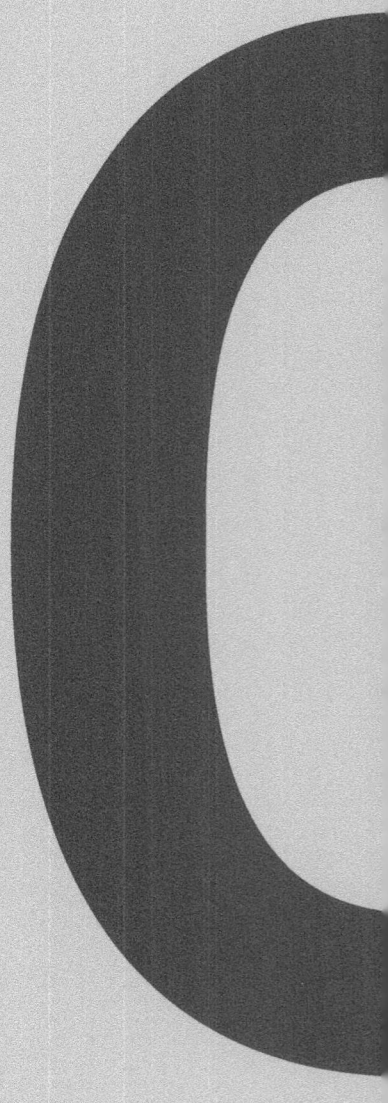

04

CHALLENGES WHEN SEEKING A STRATEGIC FIT

Different challenges and different strategies

It is a complicated challenge to create a product design that generates a long-lasting strategic fit with the user *and* the market *and* the company. Many types of knowledge, insights and ideas need to fall into place to create a perfectly long-lasting strategic fit. The challenge is also dependent on the specific situation and assignment that the design team is tasked with. If the company previously had a product with a long-lasting strategic fit, then that product exhibits essential principles that should guide the new design. Likewise, if a product within the company's portfolio has been found to have had some kind of temporary fit that now appears weak or indicates a mismatch, it may suggest very relevant information when it comes to avoiding making the same mistake in relation to the new product. Hence, different situations pose different challenges and require different strategies with regard to creating a long-lasting strategic fit.

At the beginning of a new design project, the expert design teams expend a lot of effort identifying strengths and weaknesses within the current strategic fit. They explore how the existing products in the company's portfolio fit with the user, the market and the company. It is possible that the strategic fit with the user has weakened over time. It is also possible the strategic fit with the market is based on temporary trends. And it is even possible that the strategic fit with the company appears to be weak from a long-term perspective. Every design situation or assignment poses a specific challenge with respect to creating a long-lasting strategic fit, and even if the expert design teams appear to be following the same design processes, the strategic challenges they approach and the strategies they apply will be different.

For the expert designers, the new product design becomes a means of creating a new and long-lasting strategic fit. They do so with the intention of building upon the strengths of the existing strategic fit and finding ways to overcome the weaknesses associated with it.

The expert designers analyse the current strategic fit to identify the main strategic challenge they need to solve and, accordingly, how they should apply their focus and effort during the design process. For instance, they identify the temporary fits that have resulted in potential mismatches with the user, the market and the company. This is illustrated in Figure 4.1, where the proximity of the triangles indicates whether the strategic fit is lasting (close) or weak or an outright mismatch (distanced).

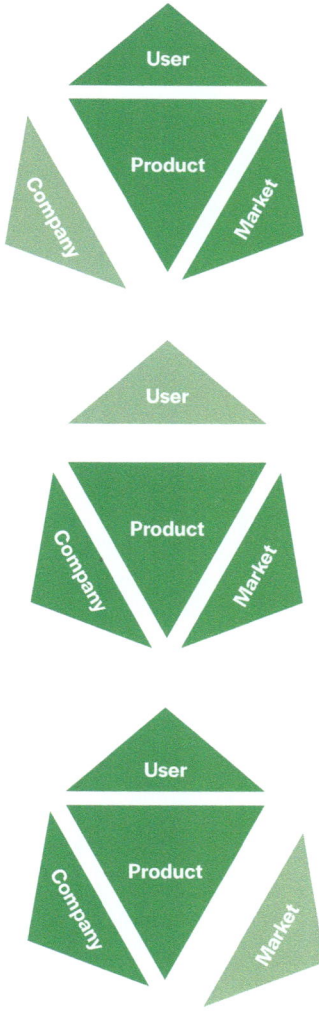

Figure 4.1: Product-company, product-user and product-market mismatches

The five strategic fit challenges

Through our research with the expert design teams, we identified five archetypical challenges related to the creation of strategic fits. Moreover, we codified how the expert designers make sense of these strategic challenges, which is defined as the way they perceive the weaknesses in the current strategic fit. The five strategic fit challenges are summarised in Figure 4.2

Strategic fit challenge 1:
Transferring a long-lasting fit

The first strategic design challenge involves designing a new product for a portfolio where previous products have been successful in creating long-lasting strategic fits and ensuring that the new product design does not disrupt this success.
The challenge is specific to the transfer of the principles associated with the present/previous products' long-lasting fits to the new product.

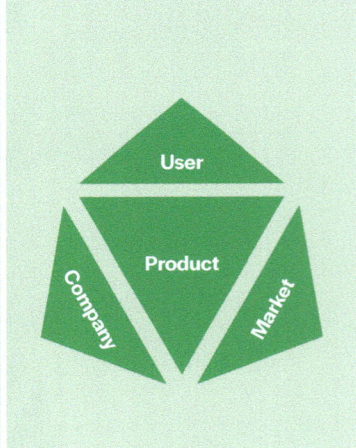

Strategic fit challenge 2:
Strengthening the product-company fit

The second strategic challenge involves designing a new product for a portfolio where previous products did not have strong fits with the company. More specifically, the challenge involves strengthening the product-company fit and avoiding the new product generating the same temporary fit with the company as some of the previous products.

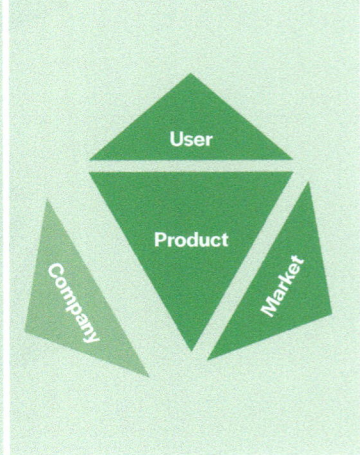

Strategic fit challenge 3:
Strengthening the product-user fit

The third strategic challenge involves designing a new product for a portfolio where previous products did not have strong fits with the user. More specifically, the challenge involves strengthening the product-company fit and avoiding the new product generating the same temporary fit with the user as some of the previous products.

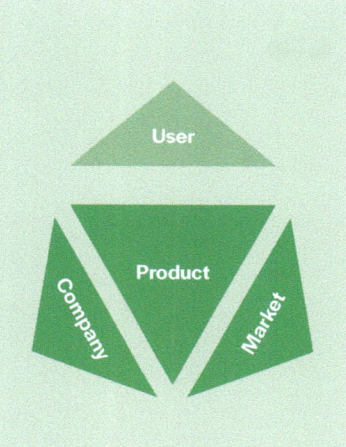

Strategic fit challenge 4:
Strengthening the product-market fit

The fourth strategic challenge involves designing a new product for a portfolio where previous products did not have strong fits with the market. More specifically, the challenge involves strengthening the product-market fit and ensuring that the new product creates a long-term competitive advantage in the market.

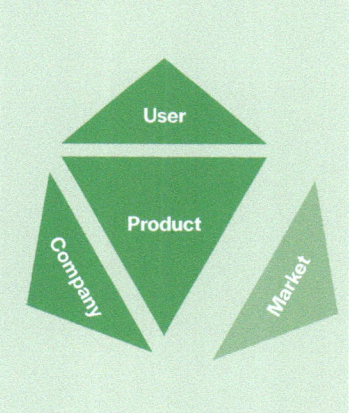

Strategic fit challenge 5:
Framing a new and lasting strategic fit

The fifth strategic challenge involves creating from scratch a long-lasting strategic fit when it is not evident from the outset what to build on and what to avoid.

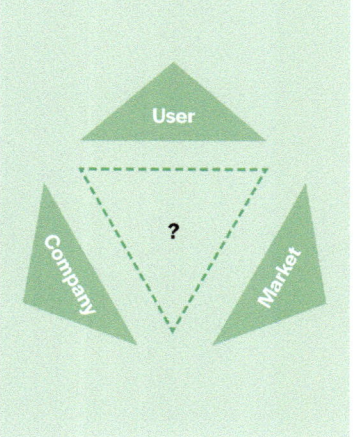

Figure 4.2: The five strategic fit challenges

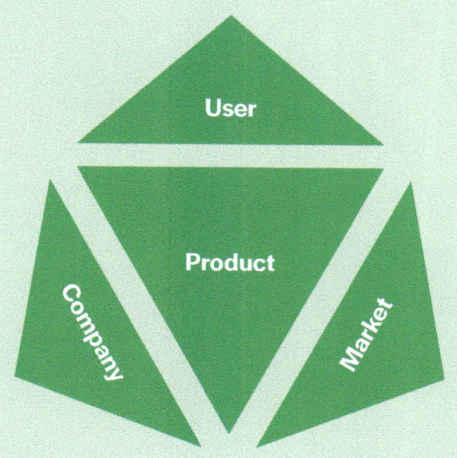

Figure 4.3: Strategic fit challenge 1:
Transferring a long-lasting strategic fit

STRATEGIC FIT CHALLENGE 1
Transferring a long-lasting strategic fit

The first strategic fit challenge involves designing a new product for a portfolio where the existing products have previously been successful in creating long-lasting strategic fits and, therefore, ensuring that the new product design does not disrupt that success. The situation may be that the company has succeeded in creating products with 'long-lasting strategic fits' and now wants to transfer the core principles of those earlier products into the new product to make it as strategically durable as possible. Another situation could be that a new product is needed to assist or supplement an already successful product within the company's product portfolio in order to strengthen the strategic fits of the company's products in general.

The fact that an existing product has proven to be successful for a long time and to have created a strategic fit between what is credible, possible and attractive for the company, as well as what is relevant to the users and the market, means that there is something to build on. Therefore, the strategic fit challenge involves transferring the current strategic fit to the new product, possibly in a renewed or updated version. This situation is illustrated in Figure 4.3.

Figure 4.4. Vipp V1 Kitchen (credit: Vipp.com)

Case

THE VIPP V1 KITCHEN

The Vipp V1 kitchen is a good example of how to transfer a long-lasting strategic fit from one product to another. It is based on the principles of the commercially successful and iconic Vipp pedal bin. In fact, Vipp had already a sound position on the market and a strong legacy due to the pedal bin, which was designed in 1939. The bin is known for its high durability and classic appearance, and many Vipp pedal bins are still in use some 40 years after being produced. Accordingly, the company had an impressive legacy that it wanted to carry on but also renew and develop with additional products in order to maintain its positioning in the market. In this sense, the kitchen was not intended to replace the bin; rather, it was conceived to assist and supplement the bin so as to both develop the legacy of the Vipp brand and strengthen the strategic fit with the users

'The first thing people often do when they move into a new house is to replace the kitchen, even though it is only five, six or seven years old and fully functional. They often want a different one, a kitchen that fit the dream of their new desired lifestyle. Our dream was to provide a kitchen design similar to the bin... People come to us after 40 years with their bin, or a second-hand one, and we change the broken pedal arm and then it could easily be used for 30 more years.'

- Designer at Vipp

and the market.

At the time, the pedal bin was sold as an accessory to new kitchens in various traditional kitchen retail stores based on "here and now" style kitchens. However, Vipp was dissatisfied with this arrangement because it did not want to be associated with 'fashionable' products, which it considered these kitchens to be. The Vipp pedal bin exhibited product longevity, which stood in stark contrast to fashionable kitchens with shorter lifetimes. In general, kitchens are said to have an average lifetime of seven years, as most kitchens are replaced even though they are still fully functional. The expert designer at Vipp explained the challenge the following way: 'The first thing people often do when they move into a new house is to replace the kitchen, even though it is only five, six or seven years old and fully functional. They often want a different one, a kitchen that fit the dream of their new desired lifestyle. Our dream was to provide a kitchen design similar to the bin... People come to us after 40 years with their bin, or a second-hand one, and we change the broken pedal arm and then it could easily be used for 30 more years.'

Vipp wanted to influence the short-term lives of kitchens and prompt people to fundamentally rethink such an approach, thereby instilling values equivalent to the values associated with the pedal bin. Hence, the company envisioned designing a kitchen that would last for decades, a kitchen that would follow you throughout a lifetime and wherever you moved to, a kitchen with high durability and classic appearance—just like the pedal bin.

Another key pattern that Vipp had observed within the kitchen industry was linked to the many options you have when buying a kitchen: 'Buying a kitchen always goes wrong… There are so many options… So you get the wrong tabletop, or the measuring is not correct. And the delivery time is delayed by three more months. Such factors are essential when it comes to the customer experience. We sought to design something that would overcome all this. For instance, if the installation was completed in one day, and then you were ready to make your Bolognese.'

As such, the design team not only focused on the physical durability of the kitchen, but on the entire customer experience from the long-term emotional perspective. They already had a product with a long-lasting strategic fit and their challenge was to apply the principles behind the pedal bin to the new kitchen.

How did Vipp overcome the above-mentioned challenges? We will come back to this case in Chapter 6.

Figure 4.5. Vipp V1 Kitchen (credit: Vipp.com)

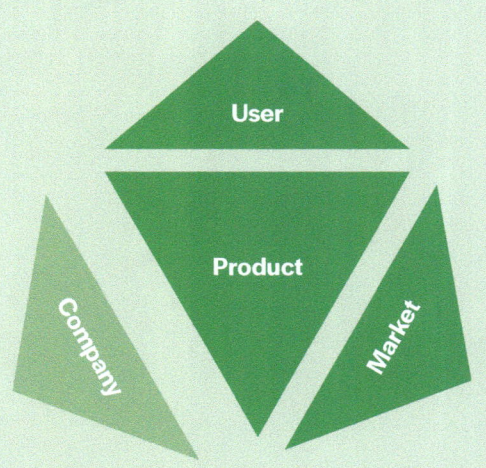

Figure 4.6: Strategic fit challenge 2:
Strengthening the product-company fit.

STRATEGIC FIT CHALLENGE 2
Strengthening the product-company fit

The second strategic fit challenge involves strengthening the product-company fit and avoiding the new product generating only a temporary fit with the company in the same way as some of the previous products in the company's portfolio did. Despite having the best intentions, companies sometimes develop products or product lines that do not fully fit with their long-term purpose and values or that end up being challenging to either produce or market, which can prove critical in the long run. Even if a product appears to be appreciated by its users and to be somewhat successful in the market, it is at risk of being discontinued if its temporary fit morphs into a mismatch with the company. Thus, when a design team is given the assignment to create a new product or new product line within an existing category, it is important to be aware of what previously resulted in a mismatch with the company in order to ensure that the new product creates a long-lasting strategic fit. The challenge embedded in this assignment is, therefore, to identify the current mismatches and reframe the new product's potential fit with the company. This situation is illustrated in Figure 4.6.

Figure 4.7. LEGO Friends (credit: iStock.com/fieldwork)

Case

LEGO FRIENDS

LEGO® Friends is the perfect example of a product line created by a design team that has engaged with the challenge of strengthening a product-company fit. In fact, the design team in question had created two previous product lines that had, in different ways, failed to create a lasting product-company fit.

In 1979, the LEGO Group introduced its first product line specifically targeted towards girls, which was known as LEGO Scala. Scala was built around a 'doll-like' main character with different accessories and play settings. In many ways, LEGO Scala imitated the most successful doll products for girls on the market. Unfortunately, Scala did not achieve the same level of success.

In 1994, the LEGO Group introduced its second product line targeted towards girls: LEGO Belville. Belville included sets featuring fairies and fantasy elements as well as sets focusing on everyday life. LEGO Belville was primarily a combination of premade landscapes and facades that would enable girls to quickly build, for example, castles and palaces in pink and purple. LEGO Belville was much more successful with girls than Scala, although compared to the LEGO Group's other product lines, it was not a real success. One of the main problems was that the LEGO bricks/components used in the Belville sets were very expensive to produce because they could only be used with Belville, not with any of the LEGO Group's other product lines. The same was true of the figures included in the Belville sets. They were much larger and more complicated to produce than the minifigures featured in most of the LEGO Group's other product lines. Belville's lack of overlap with the other product lines rendered it highly expensive to produce. Even though Belville remained on the market for a relatively long time (until 2009), it was never considered a significant success for the company.

At the beginning of the new millennium, a group of designers at the LEGO Group were tasked with resolving the challenge of engaging girls with LEGO products once and for all. The brief from company leadership was fairly simple: 'What does it take to make a LEGO experience for girls a success?'

For the design team, it was clear that previous attempts were not true LEGO experiences: 'It was one of the biggest issues before [with girls' products]... Every attempt we made at LEGO, it was not delivering a truly LEGO building experience because ... it was not on same scale of the existing LEGO brick platform, it was more unique, the product was more expensive so it was hard to have a sustainable platform for the future with the many different themes [...] So that was the driver from the beginning.'

Hence, the main challenge for the team at the LEGO Group was to provide a true LEGO experience especially for girls, one that was in line with the company's values and competencies. In other words, this round the new product should create a lasting product-company fit and still be relevant for girls.

In Chapter 7, we will review how the designers at the LEGO Group approached this challenge.

'It was one of the biggest issues before [with girls' products]... Every attempt we made at LEGO, it was not delivering a truly LEGO building experience because... it was not on same scale of the existing LEGO brick platform, it was more unique, the product was more expensive so it was hard to have a sustainable platform for the future with the many different themes [...] So that was the driver from the beginning.'

- Designer at LEGO

Figure 4.8. LEGO Friends (credit: iStock.com/Ekaterina79)

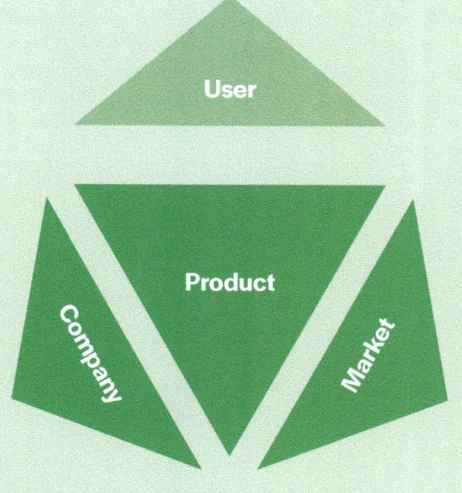

Figure 4.9: Strategic fit challenge 4:
Strengthening the product-user fit

STRATEGIC FIT CHALLENGE 3
Strengthening the product-user fit

The third strategic fit challenge involves strengthening the product-company fit and avoiding the new product creating only a temporary fit with the user in the same way as some previous products in the company's portfolio did. Sometimes, the needs, behaviours or expectations of users (or of a particular segment of users) change so much that the existing products in the portfolio are unable to fully meet them anymore. For instance, after users have new types of product experiences or interactions elsewhere, their expectations change accordingly. Although the products in the portfolio continue to have a product-user fit with some users or in some situations, it is important that the new product strengthens the product-user fit and makes it last, particularly with regard to those users with whom the fit has weakened over time.

Therefore, the challenge for the design team is to create a new product or product line that will reframe the strategic mismatch with the user. This situation is illustrated in Figure 4.9.

Figure 4.10. B&O A9 (credit: Quang Tran)

Case
THE B&O A9 SPEAKER

The case of B&O's A9 speaker is the perfect example of the third strategic fit design challenge.

At the beginning of the new millennium, B&O found itself facing something of a paradox. On the one hand, the company had built a product portfolio that was truly consistent with its mission to 'constantly question the ordinary in the search for surprising and lasting experiences'. It made products with a high degree of longevity given its high-quality brand, and it also made products that built on the company's strategic strengths and were considered attractive to the company. B&O even had a loyal and supportive customer base. On the other hand, the trouble was that the company's customer base was ageing, and B&O was struggling to resonate with the emerging new audience. Young people did not identify with either the brand or the products.

As an expert designer at B&O explained: 'When I was a young kid, I dreamed of spending all my money on a B&O music centre. Today, younger people have other preferences. They grow up with other products, such as the iPhone, Apple and the like. So how could we introduce the brand to the younger audience so that they became future B&O customers? [...] It was a critical challenge at that moment, because if we did not do anything about it, B&O would die together with the older audience.'

In the past, young people were introduced to the B&O brand through their parents. These gifts introduced them to B&O's quality and craftmanship. They also demonstrated the 'magic interactions' that B&O is so well known for, for example, the Beomaster 2400's touch panels or the Beocenter 2300's 'glass panel' that slides open when you position your hand in front of it. Such 'magical interactions' had previously been the key to connecting with the young audience and turning them into supportive and loyal future customers.

By the 1990s, B&O's products had become large sculptures of hi-fi excellence that were installed in the home in such a way that no wiring was visible. However, this was not attractive to young people and did not match the nomadic way in which many lived their lives. The products that young people iconised and spent their money on were very different to B&O's sculptures. The iPhone had just been introduced and had already resulted in new types of behaviours due to allowing music to be listened to in many different locations. Moreover, the Apple universe had created new types of expectations with respect to consumer electronics. Clearly, a mismatch existed between the highly sculptural and excellent B&O products and the products that resonated with young people.

In Chapter 8, we will review how the design team at B&O approached this challenge.

'When I was a young kid, I dreamed of spending all my money on a B&O music centre. Today, younger people have other preferences. They grow up with other products, such as the iPhone, Apple and the like. So how could we introduce the brand to the younger audience so that they became future B&O customers? [...] It was a critical challenge at that moment, because if we did not do anything about it, B&O would die together with the older audience.'

- Designer at B&O

Figure 4.11. B&O A9 (credit: Quang Tran)

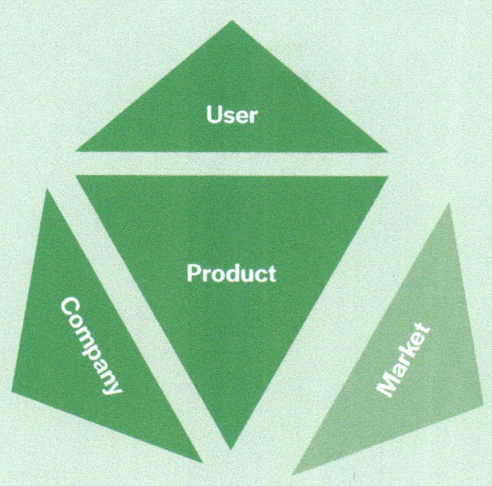

Figure 4.12: Strategic fit challenge 4:
Strengthening the product-market fit

STRATEGIC FIT CHALLENGE 4
Strengthening the product-market fit

The fourth strategic fit challenge involves strengthening the product-market fit and ensuring that the new product will be competitive in the market in the long term. It could be that, for a certain period of time, a company has provided a product with high strategic durability (e.g. due to its high quality, functionality, interaction, etc.) but that now competitors are catching up because they can provide the same features, quality level or special interactions that previously gave the company its competitive edge. This means that what were originally the products' differentiating factors have now become commodities. Nevertheless, the current products still create a fit that is credible, attractive and viable for the company, although if a new product or product update needs to be developed, it is important that it strengthens the fit with the market. This situation is illustrated in Figure 4.12.

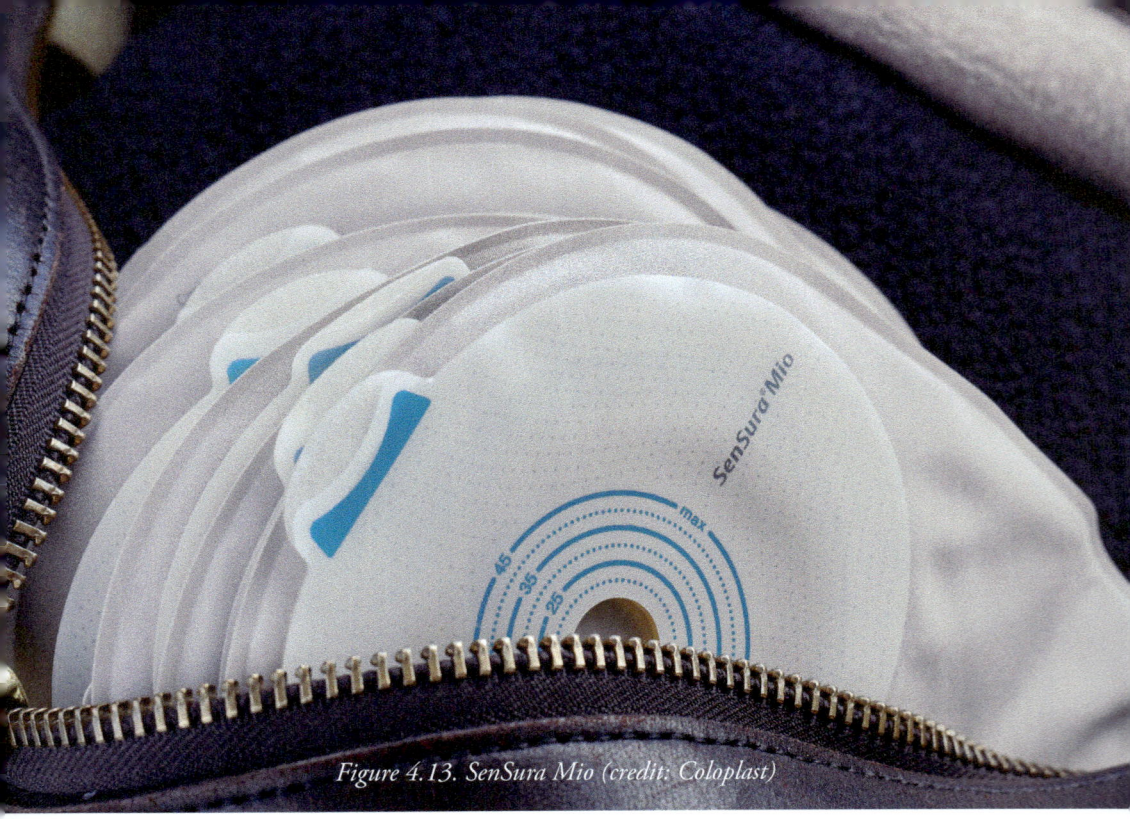

Figure 4.13. SenSura Mio (credit: Coloplast)

Case

COLOPLAST'S SENSURA MIO

Coloplast's SenSura Mio serves as an excellent example of the fourth strategic fit challenge.

Coloplast is a Danish company that was launched in 1957 with the idea of creating an adhesive ostomy bag that would minimise the risk of leakage. The aim with regard to this adhesive bag was to allow people with an ostomy the possibility to move and socialise without constantly being afraid of leakage.

Almost from the beginning, Coloplast has held a very strong position in the market, and over the years the company's focus has been on continuously optimising the bag to the point where leakage and bad smells are kept to a minimum. Coloplast has also engaged in strong collaboration with its product users, which has allowed the company to develop deep insights into what life with an ostomy is like.

However, at the beginning of the 2000s, the company found itself in a situation where its main competitors were starting to catch up. Thus, Coloplast and the expert design team had to (re)define what the company was and where it wanted to go in the future.

This was the starting point for the design of the SenSura Mio. One of the expert designers explained it in the following way: 'Our ambition was to lead the way in the market and thereby show the direction for these types of products. [...] The core thinking at Coloplast is empathy—wishing the very best for our users. In this case, we wanted to change the paradigm and the entire experience of an ostomy bag, knowing that the functionality would remain the same.'

In other words, the challenge for Coloplast was to explore and discover a new and long-lasting strategic fit with the market. In Chapter 9, we will review how the expert design team achieved this.

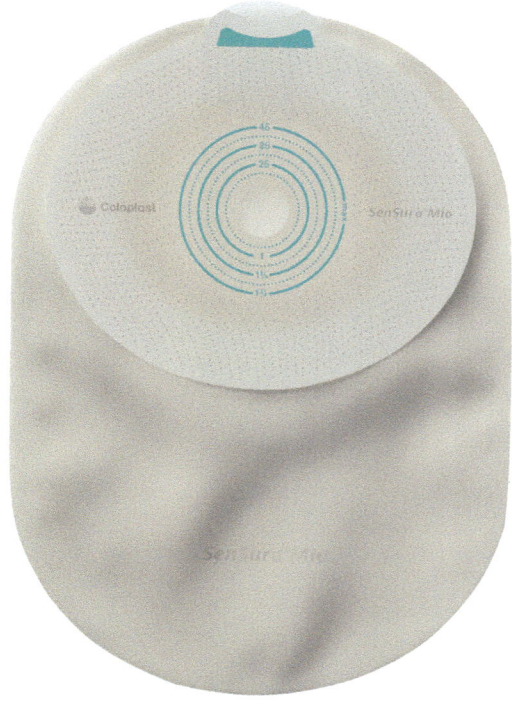

Figure 4.14. SenSura Mio (credit: Coloplast)

'Our ambition was to lead the way in the market and thereby show the direction for these types of products. [...] The core thinking at Coloplast is empathy – wishing the very best for our users. In this case, we wanted to change the paradigm and the entire experience of an ostomy bag, knowing that the functionality would remain the same.'

- Designer at Coloplast

Figure 4.15. SenSura Mio (credit: Coloplast)

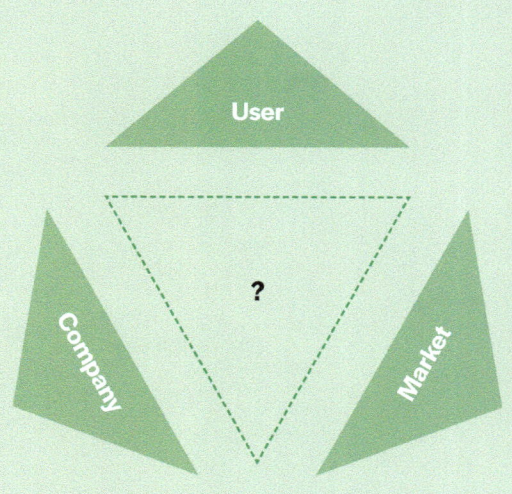

Figure 4.16: Strategic fit challenge 5:
Framing a new and lasting strategic fit

STRATEGIC FIT CHALLENGE 5
Framing a new and lasting strategic fit

The fifth strategic fit challenge involves framing a new and lasting strategic fit when it is not clear from the outset what to build on and what to avoid. Sometimes, companies that produce long-lasting products need to identify underserved markets or market segments that they can market their product to. A product's longevity means that its replacement involves a limited and slow process. Thus, the best strategy is to enter new markets with the product either in its current form or in an altered form. If there is a match between the market and the company's competencies, technologies or other resources, it is identified as a market opportunity. Therefore, the design team's challenge is to create a version of the product that targets the new market. As this process concerns a new market and segment for the company, the strategic challenge involves creating from scratch long-lasting product-user, product-market and product-company fits. This situation is illustrated in Figure 4.16.

Figure 4.17. LEGO Ninjago (credit: iStock.com/abalcazar)

Case

LEGO NINJAGO

The LEGO® Ninjago™ case is a fine example of a design team that faced the challenge of framing a new and long-lasting strategic fit for a new product.

Until 2011, when the Ninjago product line was launched, the most successful products in LEGO's portfolio were either focused on the traditional LEGO construction and creative play experience (e.g. LEGO Creator and LEGO City) or built on external intellectual properties (e.g. Star Wars). However, the aim for the Ninjago project was to create a new play universe that offered new types of play experiences for action/roleplaying children. More specifically, the design team was tasked with creating a novel project for children aged 7–9 years that focused on roleplay and featured the minifigures as a key element. 'There was not more to it than that. [...] Create a new world or a new story that

'There was not more to it than that. [...] Create a new world or a new story that LEGO had never done before. It needed to have newness in terms of not only the story but also the products. And that was it.'

- Designer at LEGO

LEGO had never done before. It needed to have newness in terms of not only the story but also the products. And that was it.' The need to create a novel story and give shape to it was the open brief that served as the starting point for Ninjago.

The key challenge that the design team at the LEGO Group had to face was the fact that they needed to create a product for children who did not typically play with construction toys such as LEGO bricks. Instead, the target group was more action-oriented, preferring roleplay to continuous construction. Hence, one of the most significant challenges for the design team

that worked on the Ninjago project was to constantly bear in mind that what they (and, for that matter, everyone else at the LEGO Group) knew about children's play with toys, particularly the LEGO Group, did not necessarily apply to the target user segment (action/roleplaying children aged 7–9). On that basis, they described the assignment as involving continuously reminding themselves that they had to start from scratch and find ways to create a lasting fit with the user, the market and the company. In Chapter 10, we provide more details as to how they did so.

Figure 4.18. LEGO Ninjago (credit: iStock.com/abalcazar)

Summary: The five strategic fit challenges

In summary, our research identified five archetypical challenges with respect to the creation of a long-lasting strategic fit. We found that expert designers expend a significant amount of effort identifying the strengths and weaknesses of the current strategic fit with the intention of building on the strengths and finding ways to overcome any weaknesses and temporalities.

The expert designers perform an analysis of the current strategic fit in order to identify the main strategic challenge they must solve and determine where they should direct their focus and efforts during the design process.

For practitioners, an overview of the specific strategic challenge they are facing may prove highly valuable due to both focusing their attention and supporting navigation through the conceptual design process.

In Chapter 13, we provide a set of **action guides** that practitioners can use to identify and qualify the strategic fit challenge embedded in a new assignment. The action guides can be used as a starting point for the creation of a strategically durable product concept.

Once you have identified your strategic fit challenge, you can apply the corresponding expert strategy. The following chapter introduces the five expert strategies for creating products with long-lasting strategic fits.

05

EXPERT STRATEGIES

Strategically durable products address users' long-term problems and needs, create long-term competitive advantages in the market, advance the long-term credibility of the company, enhance the company's strategic strengths and align with the company's long-term values, purpose and culture. When seeking to create a new product concept of this kind, it is handy to have a clear strategy for navigating the conceptual design process.

Through our research, we discovered that expert design teams dedicate a lot of effort to mapping the strengths and weaknesses of the existing strategic fit in order to identify the strategic fit challenge they need to solve when creating the new product concept. They do this because they have to use their time and effort wisely and, in particular, focus on strengthening the strategic fit and avoiding creating the same temporary fits as those associated with previous products in the portfolio. Moreover, they also do it because the new product concept could benefit from the strengths of the current strategic fit.

We have identified five expert strategies that match the five challenges reviewed in Chapter 4:

Expert strategy 1: Renewing core principles
Expert strategy 2: Leveraging objections
Expert strategy 3: Foreseeing future mismatches
Expert strategy 4: Extending product value
Expert strategy 5: Searching for hooks

In Chapters 6–10, we will review each of the strategies in great detail and illustrate them using the five cases (Vipp's V1 kitchen, B&O's A9 speaker, LEGO's Friends and Ninjago product lines and Coloplast's SenSura Mio). However, before we do so, we would first like to share some of the terminology and dynamics we observed in relation to the expert design teams' work.

In our research, we have identified three approaches that are central to the expert design teams' means of creating strategic fits: 1) renewing a strategic fit, 2) reframing a strategic fit or 3) framing a strategic fit from scratch.

Renewing a strategic fit

If there already exists a long-lasting strategic fit with the user, the market or the company (or all of them), the expert design teams must renew that strategic fit. They do so by transferring some of the core principles from an existing durable product's strategic fit to the new product, albeit in a renewed form. This process is illustrated in Figure 5.1 and will be further elaborated in Chapter 6.

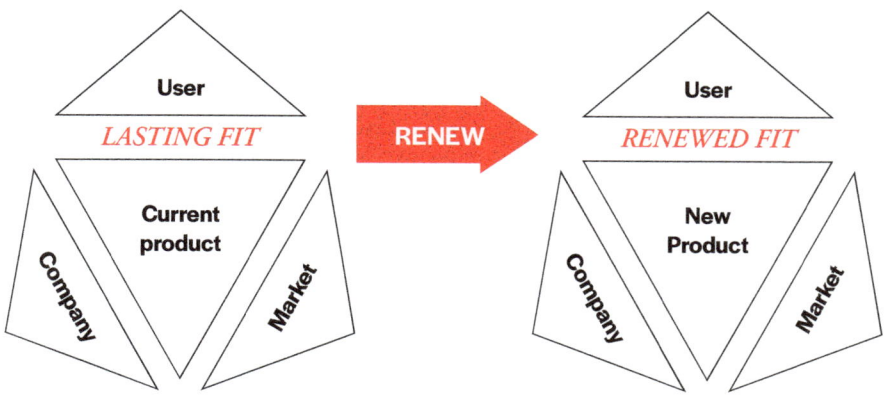

Figure 5.1: Renewing a strategic fit

Reframing a strategic fit

If a current product has a temporary fit that now appears to be a weak fit or mismatch with the user, the market or the company, the expert design teams must reframe the strategic fit. They do so by identifying the background to the product's mismatch and then using the derived knowledge as the starting point for reframing the new product concept with the aim of creating a long-lasting strategic fit. This process is illustrated in Figure 5.2 and will be further elaborated in Chapters 7–9.

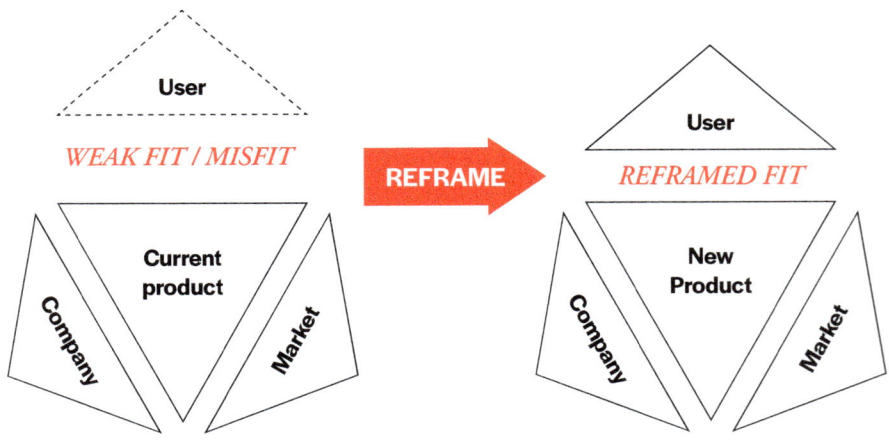

Figure 5.2: Reframing a strategic fit

Framing a strategic fit

If a previous product or any of the products currently in the product portfolio have not created a fit with the user, the market or the company, the expert design teams must create the strategic fit from scratch. They do so by identifying successful durable products that have generated valuable, meaningful and lasting strategic fits with the user, the market and the company and then using the derived knowledge as the starting point for framing a new and lasting strategic fit. This process is illustrated in Figure 5.3 and will be further elaborated in Chapter 9.

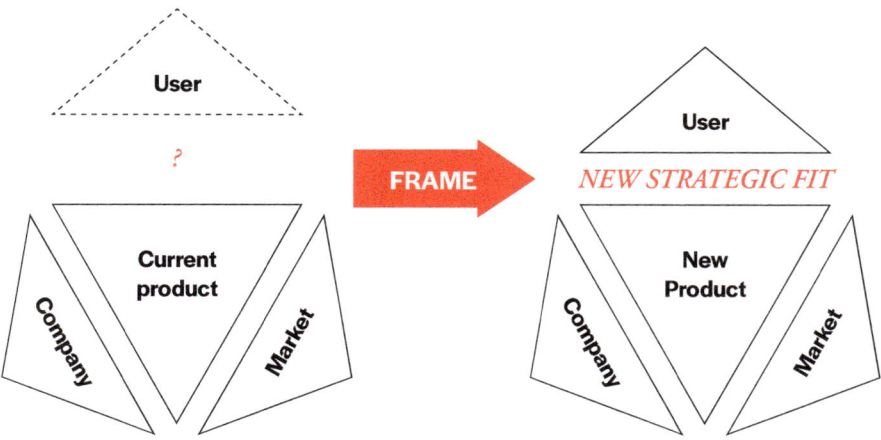

Figure 5.3: Framing a strategic fit

Expert strategies complement the design process

In the following chapters, we will work through the five different expert strategies and share with you unique insights into how expert designers approach the five strategic fit challenges described in Chapter 4. Each chapter will review a different strategy and illustrate its application using a case study.

We recommend that you read *all* of the strategies and not just the one that best matches your challenge, as we are confident that doing so will contextualise the specific strategies and provide helpful nuancing. In Chapter 13, you will find action guides for all of the strategies, which will help you to implement them in practice. You only need to read the action guide that relates to the strategy you eventually choose to implement.

EXPERT
STRATEGY 1

STRATEGIC FIT CHALLENGE 1

Transferring a long-lasting strategic fit

The first strategic design challenge involves designing a new product for a portfolio where the products have previously been successful in creating long-lasting strategic fits and ensuring that the new product design does not disrupt that success. The challenge concerns the transfer of the principles associated with the present/previous products' long-lasting fit to the new product.

EXPERT STRATEGY 1

Renewing core principles

The main focus of this strategy is on identifying the core principles that made an existing durable product fit with the user, the market and the company and then transferring them to the new product in a renewed form.

Expert strategy 1: Renewing core principles

The first expert strategy is known as 'renewing core principles'. The main focus of this strategy is on identifying the core principles that allowed the products in the existing portfolio to create lasting strategic fits and then transferring them to the new product in a renewed form. The expert designers perform a thorough analysis of the existing products' strategic fits with the user, the market and the company and then transfer the salient aspects in a renewed form to the new product. The expert designers undertake three key steps in relation to this strategy: 1) identify the core design principles behind the existing durable products, 2) renew and implement the core design principles within the new product and 3) search for ways to strengthen the strategic fit of the new product to render it long-lasting.

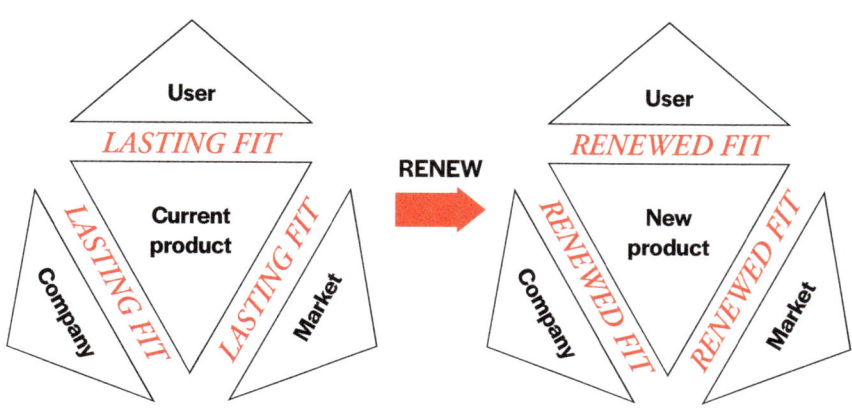

Figure 6.1: Expert strategy 1: Renewing core principles

Step 1: Identify core principles behind an existing durable product

A central part of the first expert strategy involves identifying the core principles that allowed an existing durable product to create lasting product-user, product-market and product-company fits.

First, the expert designers identify the core principles that allow the existing product to satisfy the long-term needs, wishes and desires of the user. This may entail specific long-term problems that the existing product solves particularly well, product experiences offered by the existing product that the user appreciates or means by which the existing product supports the user's behaviours or aspirations in a way that really resonates with the user.

Second, the expert designers also identify the core principles that render the existing product competitive in the market in the long term and provide it with a strong position. Here, the expert designers identify the core principles that prompt customers to select the existing product over competing products or the core principles that place the existing product in an attractive position relative to competing products.

Finally, the expert designers identify the core principles that render the existing product credible and attractive for the company. The expert designers search for ways in which the existing product builds on the company's core competencies, knowledge or strategic strengths or for ways in which the existing product sharpens or materialises the company's values and vision.

Step 2: Renew and implement the core principles

Next, the expert designers explore how best to transfer the core principles that created the existing product's strategic fit into the new product. The way in which the expert designers do this is quite individual. For some, the knowledge regarding the existing product's core principles is just intuitively implemented during the idea development and conceptualisation of the new product. The expert designers use this knowledge in a similar way to any other type of 'design knowledge' that needs to be implemented during the conceptualisation of a new product. It is a form of tacit knowledge that is intuitively implemented in ideas/concepts and prototypes when doing so makes sense.

Other designers use the core principles in a more structured way. For instance, by exploring what types of features could support the different core principles or by using the core principles as parameters for evaluating different product ideas and concepts in relation to each other.

Still, there is one thing that all the expert designers have in common, namely they focus on ensuring that all decisions and features are aligned with the identified core principles.

Step 3: Strengthening the strategic fit

During the process of creating a new product concept, the expert designers also focus on how to strengthen the strategic fit of the new product in as many ways as possible. To achieve this, they explore ways to make the product concept resonate and be relevant in the long term for the user, ways to render the product competitive in the market in the long term and ways to make the product attractive to the company and aligned with its vision in the long run. Although existing products have lasting strategic fits, the expert designers search for ways in which these fits can be strengthened, including searching for significant changes in consumer needs and behaviours or changes in the competitive landscape that need to be integrated or updated in the new product concept. In other words, even if the expert designers are building on the core principles behind existing products, they still expend significant effort in securing a lasting strategic fit between the new product concept and the user, the market and the company.

Expert example
THE VIPP V1 KITCHEN

STEP 1:
Identifying the core principles behind the Vipp pedal bin

When creating the Vipp V1 kitchen, the design team analysed the strategic fit of the iconic Vipp pedal bin. They first identified the principles that caused the pedal bin to resonate with the user. They then identified the principles that gave the bin its long-term competitive advantage and secure position in the market, and they also identified the principles that rendered the pedal bin credible and attractive for the company over many years.

The pedal bin's product-market fit

The first core principle that the design team at Vipp identified in relation to the pedal bin was termed 'The black trench coat'. The bin was designed in 1939 and remains in production in a version that is almost identical to the original version (only a few improvements have been made, such as the soft open/close mechanism). 'The black trench coat' refers to the bin's classic, long-lasting appearance and its ability to survive changing fashion cycles and tendencies over the decades. While a regular bin is something you generally hide away, the Vipp bin is different. It is displayed in many homes in a similar way to other interior design aspects. The expert designer explained why the bin has gained this unique position in the market: 'It fits into many different contexts. Durable design does not mimic a certain interior style. It is something by itself'.

Hence, the classic appearance of the bin has remained relevant and attractive from the perspective of users for decades, which has resulted in its long-term and strong positioning in the market.

Figure 6.2: The Vipp V1 kitchen. Credit: Vipp.com

The pedal bin's product-user fit

The second principle that the expert design team identified in relation to the pedal bin was termed 'It's a tool'. The bin was originally crafted by a young metalworker, Holger Nielsen, for use in his wife's hairdressing salon. It was intended for everyday use and it was made easier to access when you had something in your hands. It was a tool designed to ease Marie Nielsen's work at the salon.

The tool references are embedded in the details and interactions of the bin. As the designer explained: 'The bin was designed as a tool for a hairdressing salon. Most of the design comes from the interactions and functionality. It has ears so you can move it. The shape is wider in the bottom for it to stand more stable on the floor.

Figure 6.3: The Vipp pedal bin. Credit: Vipp.com

It has a pedal so it is possible to operate the lid with full hands.' This tool feeling represents an important aspect of the entire product experience of the bin, and it has proven to resonate well with Vipp's customers. During its first 50 years, the bin was primarily for professional use, although today it also targets the private market, where it is considered attractive and relevant to many different types of users.

STEP 2:
Renewing and implementing the core principles in the new kitchen

The design team at Vipp also explored how to transfer the identified core principles—both 'The black trench coat' and 'It's a tool'—from the pedal bin to the new Vipp V1 kitchen.

Core principle: The black trench coat of kitchens

During the process of designing the new Vipp kitchen, the design team engaged in continuous efforts to transfer the 'The black trench coat' principle from the bin to the kitchen. This principle mainly referred to the appearance of the kitchen: 'A main aim of the design process was to see if we could catch something that would never go out of fashion. The design should be as classic as a black trench coat.' The designers implemented this principle in a number of ways. First, in contrast to fashionable kitchens, they found that the kitchen should be an example of long-lasting furniture that fit into many different places due to its classic and neutral appearance. For instance, there should only be a black or white version. In addition to the long-lasting appearance, the kitchen should communicate physical durability through the utilised materials and its high quality, similar to what the bin was known for. This was achieved by integrating the bin's main materials in various forms, such as a stainless-steel tabletop. This first principle is described in Table 6.4.

Core principle: 'It's a tool'

The design team at Vipp also made an effort to implement the 'tool' experience in relation to the new kitchen, mainly by ensuring that all features and

Table 6.4: Design principle 1

design decisions supported this aspect. Similar to how the bin had its roots in the professional market, the Vipp kitchen was also inspired by professional industrial kitchens. As the expert designer explained: 'The bin was not intended to be what we today would call a typical design object. It was a tool for Marie's salon. It would operate 20–30 times a day. Essentially, we like the functional tool thinking. The translation of a professional industrial kitchen into a domestic kitchen is not about making all the parts in stainless steel and ensuring they are super-optimised for intensive cleaning; rather, it is the essence we take from it. For instance, it has legs, the tabletop is made from stainless steel, and then we add the elements that make it fit into the living room.' The designer further explained how they worked with the tool experience: 'We paid great attention to the interaction and touch points by additional detailing. There are extra details where you touch it. For instance, the shape of the gas buttons… they have been milled in such a way that… they are almost machinelike and have a very nice feeling when you touch, feel and turn it… The same goes for the front handles in extruded and bended aluminium that have a

Strategic fit with the user:

The tool experience mimics a well-functioning product that has resonated well with Vipp's customers because it has proved to be highly durable in terms of everyday usage and they love using it.

Solutions in the pedal bin	Solutions in the new kitchen
• It has ears to move it • The shape is broader at the bottom so it feels more stable • A pedal is included to ease opening with something in the hands • There is an inner bin to ease cleaning	• Machinelike details/interaction points (e.g. the milled gas regulator buttons) • References to the classic industrial kitchen workspace (e.g. stainless steel table top, standing on legs)

Table 6.5: Design principle 2

recessed rubber component on the invisible side to enhance the grip and user experience'.

The tool experience was an important strategy for the design team in terms of increasing the chance of an emotional connection being built. For instance, the feeling when turning the gas button should provoke a sense of quality and pleasure, similar to how people love the sound of a closing BMW door.

One of the most important goals for Vipp was to make people love using the new product on a daily basis as opposed to design features that are solely visual. The 'It's a tool' principle is described in Table 6.5.

STEP 3:
Strengthening the strategic fit

The design team at Vipp also explored whether there were areas where the kitchen's strategic fit needed to be strengthened to ensure its long-term fit with the user, the market and the company.

Strengthening the fit with the market

During the process of designing the kitchen, the design team had looked at the present market for kitchens and realised that there were several issues needing to be addressed in the Vipp version of a kitchen. First, the aim was to differentiate the Vipp kitchen from the fashionable kitchens by designing a durable kitchen, both in terms of physical durability (i.e. durable materials and quality) and emotional durability (i.e. people would keep and love it for many years). Second, they identified the potential to generate a competitive advantage and differentiate from the many customisation possibilities offered by existing kitchen solutions. According to the design team, the many options could easily prove overwhelming for the customer and increase the complexity and risk associated with the design process. Building on the core competencies within the company, these issues should be eliminated in the Vipp kitchen and, therefore, become the key differentiators with regard to kitchens currently on the market.

Design principle: 'The nomad kitchen'

The 'nomad' idea was motivated by the design team's experience of well-functioning kitchens

---

Design principle 3:

THE NOMAD KITCHEN

Strategic fit with the market (differentiation):
The kitchen should be an investment for the customer, rather than for the particular house or apartment (take it with you when you move, a kitchen to last a lifetime).

Solutions in the pedal bin	Solutions in the new kitchen
• The pedal bin intends to follow you for a lifetime; design for generations	• Free-standing modules that can be repositioned and legs can be adjusted

Table 6.6: Design principle 3

being replaced after just a short lifetime. The main idea was to start seeing the kitchen as a piece of furniture you can potentially take with you when you move: 'a kitchen to last you a lifetime'. Hence, the Vipp kitchen was not designed to be an investment for a particular house or apartment because it was customised and fixed to the floor and walls. Instead, it was an investment for the customer who could take it to a new home. 'The nomad kitchen' consists of free-standing modules that can be repositioned and have legs that can be readjusted. Another advantage, perhaps even more value creating, of the nomad solution having only three types of modules concerned the planning and installation process: 'When a customer comes to us with a floor plan, it takes only five minutes to find a good solution. We engaged with all of the challenges we found in existing solutions and so solved and eliminated the problems through the design. For instance, cleaning is made easier by the legs, which also make the installation more operational, meaning that the kitchen can be installed by locals in South Africa, for instance.'

Thus, as the designer explained, the nomad idea not only made the installation process easier

---

Design principle 4:

FORD T-TYPE CHOICES

Strategic fit with the market (differentiation):

The market for the online sale of kitchens had started to mature. The limited options ease the planning and installation process, which optimises the potential for online sales and, therefore, the potential to be present worldwide.

Solutions in the pedal bin	Solutions in the new kitchen
• Limited customisation possibilities (only the colour varies)	• All appliances are pre-chosen; the only choice to make is between a gas range and an induction range

Table 6.7: Design principle 4

and more operational, it also created an opportunity for Vipp to enter the international kitchen market. 'The nomad kitchen' design principle is summarised in Table 6.6.

Design principle: 'Ford T-type choices'

Finally, the expert design team adopted the 'Ford T-type choices' design principle (Table 6.7) in relation to the process of buying a kitchen. In the case of most kitchens, the customer is overwhelmed with options such as materials, colours, white goods, sinks and fixtures, which means that the planning and buying process is often lengthy. The 'Ford T-type choices' principle refers to Henry Ford's famous quote, 'Any colour – so long as it's black', which was the main thinking during the process. The aim was to create a kitchen with as few options and customisation possibilities as possible, in contrast to the competing kitchens on the market. As the expert designer explained: 'You don't have different options of white goods. We have made all choices. You can select between a gas range and an induction range, but that's it. The tabletop is stainless steel. We don't offer any options. Instead, we take pains to find the best solutions and then we hope that enough people will like the kitchen. […] It's very much take it or leave it.'

This approach also embedded the potential for a new sales channel that had just started to mature in the kitchen industry. The limited options (and, consequently, the simple planning and installation process) were extremely well suited to an online sales channel. Indeed, this was a great opportunity for Vipp to be present as a brand in the global context, not only for the kitchen but also for the pedal bin.

Expert strategy 1 in practice

Although the example of the Vipp V1 kitchen may seem unique due to the length of time that passed between the pedal bin being created in 1939 and the creation of the new Vipp kitchen, the challenge of creating a new product to assist or complement already successful products in the portfolio is relatively common. In fact, we have seen several examples of expert design teams renewing a lasting strategic fit and transferring core design principles from an original product to a new one.

The expert designers we engaged with typically had deep insight into the product, the product categories and the market, which enabled them to easily understand the core principles behind the original product's strategic fit. When viewed from the outside, the 'identification' of the core principles might seem fairly straightforward. However, based on our practical experience, we know that it can prove to be a very slow and challenging process. Moreover, the design team might not always have the level of experience or insight concerning the product, user, market and/or company that is required.

In the action guide for strategy 1 (see Chapter 13), we suggest a number of actions that we know from experience can provide access to the knowledge required to understand a product's core principles. However, we must underline that the work involving design principles is a skill that takes time to master.

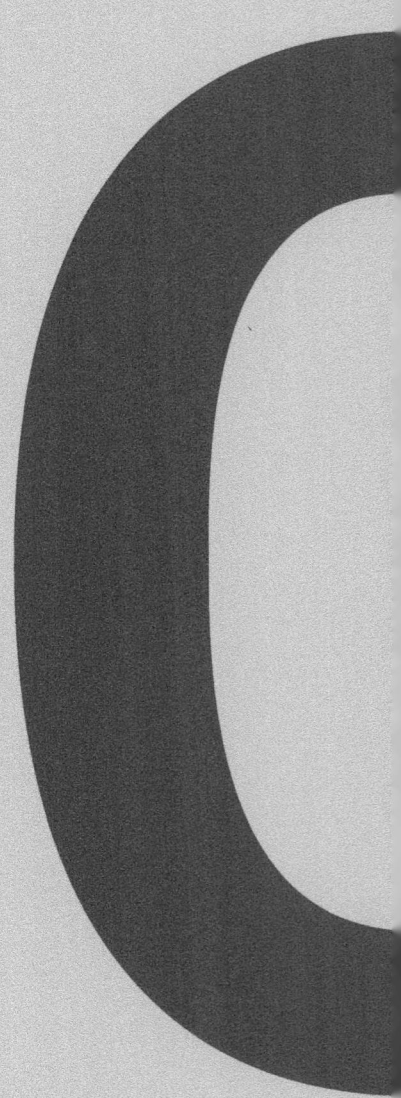

07

EXPERT STRATEGY 2

STRATEGIC FIT CHALLENGE 2

Strengthening the product-company fit

The second strategic challenge involves designing a new product for a portfolio where the previous product does not have a strong fit with the company. More specifically, the challenge involves strengthening the product-company fit and avoiding the new product concept creating the same temporary fit with the company as some previous products in the portfolio.

EXPERT STRATEGY 2

Leveraging objections

The main focus of this strategy is on identifying the reasons for the product's mismatch with the company and then using the derived knowledge as the starting point for reframing the new product.

Expert strategy 2: Leveraging objections

The second expert strategy is termed 'leveraging objections'. The main focus of this strategy is on identifying the reasons why some of the products in the current portfolio have created weak strategic fits with the company and then using the derived knowledge as the starting point for reframing the new product. The expert designers pay attention to all of the objections that have been raised against the products from both inside and outside the company and then identify the reasons behind the products' weak fits or mismatches. In this way, they leverage their knowledge of the temporary fits to reframe the new product and create a long-lasting strategic fit.

The expert designers follow three key steps in relation to this strategy: 1) identify objections and strategic mismatches, 2) reframe the strategic mismatches and 3) test and align the new strategic fit.

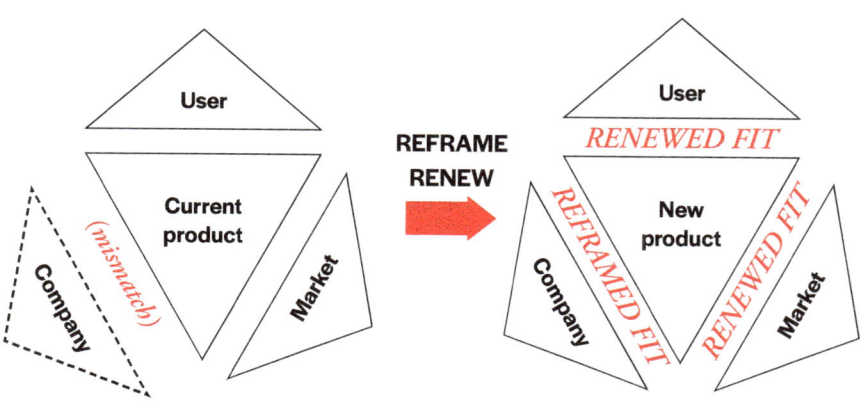

Figure 7.1: Expert strategy 2: Leveraging objections

Step 1: Identify objections and strategic mismatches

At the beginning of the design process, the expert designers search for all potential objections to the unfitting product. They do so in order to identify the underlying reasons for the product's lack of success and its strategic mismatch with the company.

Internal and external objections to the unfitting product

First, the expert designers identify the temporary aspects that rendered the current product unattractive to the company. The designers pay attention to the objections raised against the product by different departments and engage with the harshest critics of the product inside the company. Those critics provide insights into the features, aspects or 'design principles' of the product that have created a weak fit or mismatch with the company over time. Sometimes, it is also possible for the expert design team to compare the unfitting product to some of the company's successful and long-lasting products. The aim is to identify what the unfitting product is missing as well as how the durable products create lasting strategic fits with the company (e.g. by integrating the company's strategic strengths and core competencies or by being aligned with the company's values, purpose and culture).

Another possible approach for the expert design team involves exploring why the current product no longer seems credible on the market. The designers talk to sales managers, marketers and other individuals who are actively engaged in promoting or selling the product in order to identify external objections to it (e.g. the reasons why they think the product is not successful or easy to sell). They do so in an effort to understand the underlying design principles behind the product that render it less credible on the market. The expert designers also compare the unfitting product to some of the most successful and long-lasting products within the same category currently on the market or, to be more precise, they identify the 'design principles' that have given these products long-term competitive advantages or unique market positions in order to identify what the unfitting product is lacking.

Step 2: Reframing the strategic fit

During the design process, the expert designers use the knowledge concerning the unfitting product's mismatch with the company as the starting point for reframing the new product's strategic fit. They do so by focusing on the reason for the mismatch and then suggesting ways to overcome it.

They also take inspiration from products that have succeeded in creating long-lasting fits with either the market or the company in order to identify potential lasting strategic fits. The inspirational products are selected on the basis that they are highly successful and durable, generate lasting competitive advantages or prove particularly attractive to the company. During this process, the expert designers sample successful design principles from many different products, typically across different types of product categories. None of the studied products have created the same strategic fit that the design team needs to create for the new product, although they have all solved a smaller part of the strategic fit challenge and, therefore, are highly relevant to sample 'design principles' from. For example, one product might be competitive in the market in the long term while other products might have lasting product-user fits.

The expert design team uses this sampling of design principles from different inspirational products along with insights into the unfitting product's strategic mismatch as the starting point for creating the new product and reframing its strategic fit.

Step 3: Testing and aligning the new strategic fit

When creating the new product, the expert designers also focus their energy and effort on creating and aligning the strategic fit of the new product in as many ways as possible. To achieve this, they explore new ways of making the new product meet the long-lasting needs of the user, identify ways to make it competitive in the market in the long term and find new ways to make it attractive and align with the company's vision in the long term. Moreover, as the expert designers are replacing an unsuccessful (temporary) strategic fit, they need to constantly test the strength of the new product's strategic fit to see if it is lasting. They do so by continuously testing their product ideas and concepts with users as well as by presenting the concept to key sales personnel, marketers and members of the strategic management team. In other words, the expert designers continue to look for potential objections from the user, the market and the company throughout the process of creating the new product to ensure that it creates a long-lasting strategic fit.

Expert example
LEGO FRIENDS

STEP 1:
Identify objections and strategic mismatches

One of the interesting things that the design team at the LEGO Group did after being given the brief to 'Make a successful LEGO experience for girls!' was to explore why LEGO's previous attempts at creating girls' products, namely LEGO Scala and LEGO Belville, had not created lasting strategic fits with the company.

Scala's and Belville's mismatches with the LEGO company

First, the designers explored the unfitting products' mismatches with LEGO as a company. This process was initiated by paying attention to some of the key objections raised by different LEGO employees concerning both Scala and Belville.

Many of the objections from within the company focused on the fact that both Scala and Belville were very expensive to produce because they did not use the same moulding forms as LEGO's other product lines. One of the core strengths of the LEGO Group's production process is that the company reuses moulds across product lines to both lower production costs and increase creativity among designers (making them use the same elements in many different ways across themes), although this advantage was not implemented with regard to either Scala or Belville.

Moreover, it was found that both Belville and Scala were criticised for not being 'true' LEGO products. Thus, the design team started by determining the reason for this. It became clear that there was a mismatch between the LEGO Group's mission and values and the type of play that both Belville and Scala offered. The LEGO Group's mission is 'to inspire and develop the

Figure 7.2: LEGO Friends. Credit: iStock.com/cjmacer

builders of tomorrow'. The company believes that it is important to give children both creative confidence and creative skills, so that when they grow up, they can change the world. Therefore, many LEGO products are built on the 'pride of creation' design principle. The main idea behind this principle is that when children have a successful creation experience and make something they can be proud of, it encourages them to initiate new and more advanced creations, thereby building up their creative skills and confidence. Yet, when the designers looked at LEGO's previous products for girls, they did not see much evidence of 'pride of creation'. With regard to Scala, there was nothing to build at all, while in terms of Belville, the building experience was limited to placing facades on a pre-made platform of hills and slopes. Hence, to create a successful LEGO experience for girls, the designers asked themselves: 'What does it take to make a true LEGO experience for girls?'

Scala's and Belville's mismatches with the market

The design team at the LEGO Group also explored why Belville and Scala did not seem credible in the market. Here, they compared Belville and Scala to some of the most successful and lasting products on the girls' toy market at the time, especially those toys that girls liked and had on their wish lists year after year. However, rather than looking at features to benchmark or patterns to imitate, the designers focused on identifying the key design principles that created these successful products' fits with the market.

One of the key findings of the comparison was that the most successful products on the market were much better at creating play with full story characters. Based on their research, they learned that characters are important: 'Characters are the entry point… the appeal, the detail.'

By contrast, both Belville and Scala featured only limited characters. Hence, the designers realised that they needed to create a much more extended set of characters for the new product. During their research, they discovered that they needed to create a full persona (not just a hero and a mission), that is, a full character with a house, a pet, different passions, specific clothing, etc., for girls to see themselves in. 'Social play is a big thing, a big thing for girls especially. So, characters are very important to ignite in the play.'

STEP 2:
Reframing the strategic mismatches

During the design process, the design team at the LEGO Group used their knowledge of Belville's and Scala's mismatches with the company as the starting point for creating the new product. They did so by focusing on

---

Design principle 1:

PRIDE OF CREATION

Strategic fit with the company:

Provide girls with the creative confidence and creative skills to change the world—in line with the LEGO Group's mission 'to inspire and develop the builders of tomorrow'

Move away from:	Be like:
LEGO Scala/Belville	Core LEGO product lines
• No or limited creation	• Freedom to create
• No play loops	• Small wins
• No small wins	• Strong play loops

Table 7.3: Design principle 1

FRIENDSHIP - ALL ARE EQUAL

Strategic fit with the company and credibility on the market:

Values of equality and tolerance.
Differentiates from most successful girls' products that focus on one main character.

Move away from:	Be like:
• Belville/Scala's limited characters • One character in front and a few supportive characters	• Friendship • A group of equal friends • The possibility for children to see themselves in one or more of the 'characters'

Table 7.4: Design principle 2

the reasons for the mismatches and then suggesting ways to overcome them or reframe the new product in such a way that would create a long-lasting strategic fit.

Reframing the fit with company values

As mentioned above, the design team had decided that their aim was not just to create a successful LEGO experience for girls, but to create a 'true' LEGO experience for girls. They initiated the work towards this ambition by exploring some of LEGO's most successful and lasting product lines (including LEGO Creator and LEGO City) and identifying the 'design principles' behind the products that successfully aligned with the company's values, purpose and culture. The most important of these was the 'pride of creation' design principle. Through strong play loops, small wins and other means, it provides children with a successful experience of creating an outcome they can be proud of, which encourages them to initiate new and more advanced creations, thereby building up their creative skills and confidence. The integration of the 'pride of creation' design principle is explained in Table 7.3.

Based on research concerning the most successful products on the market, it was clear that the new product needed more extensive characters to create a lasting strategic fit. Still, the design team also had the ambition of creating a strong position for the new product and differentiating it from the other products on the market. During the exploration of the successful products on the girls' toy market, the designers had seen that most products for girls were built around one lead character and a number of supporting characters. However, the design team found that this central positioning of a single character resonated very poorly with some of the values at the LEGO Group, which wished to celebrate all children and not just the person in front. These values also offered an opportunity to create a play experience for girls based on the principles of equality and tolerance.

The designers came up with the idea of creating a group of friends with different competencies and interests who are all equally highlighted and tolerant of each other's differences.

Each character would have a unique personality, interests and competencies that the girls playing with LEGO Friends could recognise in themselves. The aim was to

Design principle 3:

REUSE PRODUCT PLATFORMS ACROSS PRODUCT LINES

Strategic fit with the company:

Integrate one of the core strengths of the LEGO Group's production to secure long-term attractiveness for the company

Move away from:	**Be like:**
• LEGO Scala/Belville, which have many unique elements that are expensive to produce	• Core LEGO products that reuse existing elements to lower production costs

Table 7.5: Design principle 3

help girls to appreciate different competencies and interests. LEGO Friends was designed to create a 'safe' environment where children can play without feeling insecure even if they are not similar to the 'girl in front' and where they can find 'power' in what they are and the competencies they possess. The new and lasting strategic fit with the market is summarised by the 'Friendship - All are equal' design principle (see Table 7.4).

Reframing the fit with the company's core competencies

Finally, the designers also adopted a design principle that can be seen in relation to many successful LEGO products, namely the 'Reuse product platforms across product lines' principle (Table 7.5), when creating the new LEGO Friends product line. This was done to ensure that the new product line would have the same production advantages as most other core LEGO products and, therefore, be attractive to the company in the long run.

The design team aspired to reuse as many of the moulding forms from other product lines as possible to lower production costs. In the end, more than 90% of the LEGO Friends parts were made from existing elements. The only pieces that were unique to LEGO Friends were the main characters and some new accessories that supported character creation. During the design process, the designers found it impossible to place 'enough character' into the LEGO minifigures and so decided to allow the characters to be uniquely produced. This was important because the success of the characters was essential to the new product's lasting fit with the market.

STEP 3:
Test and align the new strategic fit

During the creation of the LEGO Friends product line, the design team constantly tested the strength of the new product's strategic fit. They tested their product ideas and concepts with users and other key stakeholders in order to identify any potential objections that could limit the new product's fit with the user, the market and the company as well as to ensure that internal stakeholders were informed about the project's findings.

Table 7.6: Design principle 4

'It has been an end-to-end approach, you know, talking to production, our stakeholders in the company, even top management. I called it "take leadership by the hand". [...] We had a lot of check-ins with leadership, to share with them what we were learning, what insights we were getting, and what decisions we were making based on those insights.'

Through the designers' research, it became clear that the new product needed to support girls' appreciation of 'many accessories rich in details' and invite them to set up the play, as these are essential aspects of many girls' way of playing. The most successful and lasting products on the market supported girls in this way by providing all the accessories and details needed to create a full character and multiple stories for different roleplays (e.g. houses, cars, pets, sports equipment, etc.). Inspired by these insights, the design team created the 'Layers of colour and detail' design principle to ensure that the new product would allow for such play, which should ensure a lasting product-market fit (see Table 7.6).

The LEGO Friends product line was introduced in 2012. It is based on the story of five friends—Stephanie, Mia, Andrea, Emma and Olivia— and their life in Heartlake, a suburban community.

Expert strategy 2 in practice

Although the example of LEGO Friends may seem rather straightforward based on the case description, the many prior attempts to create a successful and long-lasting LEGO product for the girls' market indicate that solving this challenge was no trivial matter. Reframing a strategic mismatch with the market and the company calls for significant attention to objections and mismatches. During our research, we have seen a number of examples where expert design teams have proved excellent at identifying objections and using inspirational products as the basis for reframing new products. However, we have also seen the opposite.

We acknowledge that it takes time and expertise to identify and notice objections. First, some objections will not emerge through a structured process; rather, they have to be overheard during discussions or critiques. Otherwise, such objections typically remain tacit. Second, it takes expertise to not only identify objections but take action regarding them and reframe objections and mismatches into a new product with a lasting fit. Finally, it is also important to acknowledge that identifying the core design principles associated with inspirational products and adopting them in a meaningful way in relation to a new product without imitating or directly copying takes time and expertise.

In the action guide for strategy 2 (see Chapter 13), we have established a process for identifying objections and inspirational products along with the core principles associated with them that can be used to support the process of creating a product with lasting product-market and product-company fits.

Expert strategy 3: Foreseeing future mismatches
Expert example: B&O A9
Expert strategy 3 in practice

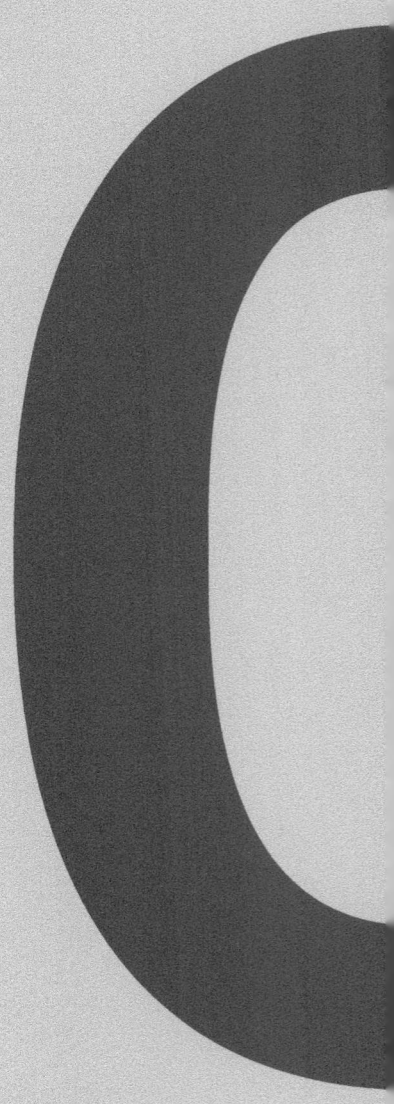

08

EXPERT
STRATEGY 3

STRATEGIC FIT CHALLENGE 3

Strengthening the product-user fit

The third strategic challenge involves designing a new product for a portfolio where previous products did not have a strong fit with the user. More specifically, the challenge involves strengthening the product-company fit and avoiding the new product creating the same kind of temporary fit with the user as some previous products in the portfolio.

EXPERT STRATEGY 3

Foreseeing future mismatches

The main focus of this strategy is on identifying potential mismatches between what the user will find attractive in the future and what the company is currently offering and then using the derived knowledge as the starting point for reframing and renewing the strategic fit.

Expert strategy 3: Foreseeing future mismatches

The third expert strategy is termed 'foreseeing future mismatches' and its main focus is on identifying potential mismatches between what the user will find attractive in the future and what the company is currently offering and then using the derived knowledge as the starting point for reframing and renewing the strategic fit. The focus on future mismatches ensures that the new product's strategic fit will be long-lasting, not temporary.

During the process associated with this strategy, the expert designers pay attention to emerging changes in users' needs, behaviours and expectations, which might lead to a strategic mismatch between users and the company's products in the future may. Based on their findings, the expert designers create a new product concept that explains how to address these new and emerging needs, behaviours and expectations on the part of users and, therefore, ensure the company's future relevance from a long-term perspective. When pursuing this strategy, the expert designers follow three key steps: 1) identify emerging changes in users' needs, behaviours and expectations, 2) reframe the strategic fit with users and 3) renew the strategic fit with both the company and the market.

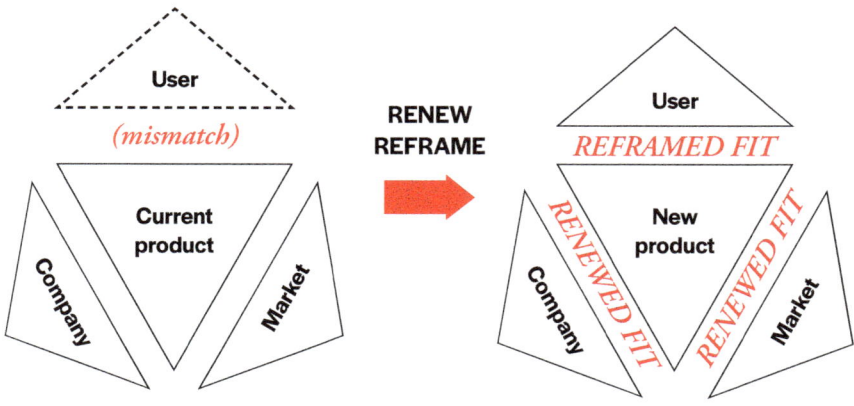

Figure 8.1: Expert strategy 3: Foreseeing future mismatches

Step 1: Identifying emerging changes in users' needs, behaviours and expectations

At the beginning of the design process, the expert design team pays attention to emerging changes in users' needs, behaviours and expectations. In fact, they pay particular attention to both first-movers' behaviour and young people's behaviour, and they research changes in technology, lifestyle and society. They do so to identify the temporary aspects of the current product-user fit that need to be addressed in the new product.

Furthermore, they identify products that have either initiated the observed behavioural changes or been particularly good at capturing such changes. These inspirational products typically come from many different product categories and various industries. Yet, what these inspirational products have in common is the fact that they are all emotionally durable products that resonate well with the exact user segments the design team is targeting with their new product and, at the same time, have been able to drive or capture key changes in users' behaviours, needs or expectations. The expert design team identifies the principles behind the products that are central to capturing or driving the new and emerging changes.

Step 2: Reframing the strategic fit with users

Next, the expert designers use the knowledge they have developed regarding the emerging changes to identify the temporary aspects of current products' strategic fits that could lead to a future product-user mismatch. Based on their insights concerning changes in users' needs and behaviours as well as the products that drive such changes, the expert designers identify potential mismatches between what users will find attractive in the future (given the emerging changes) and the products in the company's current portfolio. With these mismatches in mind, the expert designers determine why the company's existing products will not resonate with users in the future as well as why they will not meet users' future needs and expectations.

In light of the potential future mismatches, the expert designers reframe the direction for the new product or new product line.

More specifically, they use the design principles associated with the inspirational products (i.e. those that have been able to drive or capture change) as the inspiration for reframing the new product's strategic fit with the user, thereby creating a new design direction for the new product. Sometimes, the expert designers may directly transfer design principles from the inspirational products to the new product design.

Step 3: Renewing the strategic fit with the market and the company

During the design process, the expert designers ensure that the new product will have a lasting strategic fit with both the market and the company.

First, they identify the core products made by the company, including products that have lasted for a long time, products that have been particularly successful and products that the company is particularly known for. Second, they identify how those products create long-lasting strategic fits with the market and the company. In this way, the expert design team samples the successful and long-lasting design principles that have been applied across the products in the company's current portfolio. Next, they transfer the core design principles to the new product in a renewed form.

Expert example
B&O A9

STEP 1:
Identifying emerging changes in the younger audience

The first thing the expert design team at B&O did was to pay attention to emerging changes in users' needs, behaviours, and expectations. Even though B&O had a loyal and engaged customer base, the expert designers were eager to understand why their products had started to become less attractive to the younger audience. A key finding was that young people moved a lot and so did not live in the same apartment or house for a long time. Second, the expert designers looked into the new types of consumer electronics that were being marketed at the time, particularly Apple's products. These products supported high mobility and a life 'on the go'. Indeed, new consumer electronics such as phones and laptops could be used almost everywhere, and moving such products around

in everyday life and from one apartment to another was very simple.

This understanding of consumer electronics stood in stark contrast to the path B&O had been following. The company had not focused on easy mobility; rather, it had focused on making its products as integrated into the house as possible in order to avoid the need for ugly wiring. Beolink is perhaps the most eye-catching example of this approach. BeoLink made it possible to connect all B&O products in the home and have them play simultaneously. Also lights, curtains, etc. could be controlled by a B&O remote. This was most often done as part of a redecoration project or a new building to hide all the cabling in the walls, floors and ceilings. The only problem was that the use of such a successful installation system only makes sense if you intend to stay in the same house for a long time. Thus, the expert designers identified a mismatch between the young people's mobile way of living and the highly integrated and installation-intensive nature of B&O's products.

Next, as they knew that fitting hi-fi speakers into the home was a key challenge they had to solve in order to create a strong strategic fit with users, the expert design team started to explore changes in the way young people use and decorate their homes. Here, they discovered that the 'living-room chair' had become a key element of Scandinavian interior design. Young families no longer invested in large sculptures of hifi-excellence. Instead the money was spent on furniture, often a desired classic, that could easily be moved from one home to the next. This was quite different from the approach of previous generations, where

Figure 8.2: B&O A9. Credit: Quang Tran

**Young people's mobile
way of living**

**B&O's highly integrated and
installation-intensive products**

MISMATCH

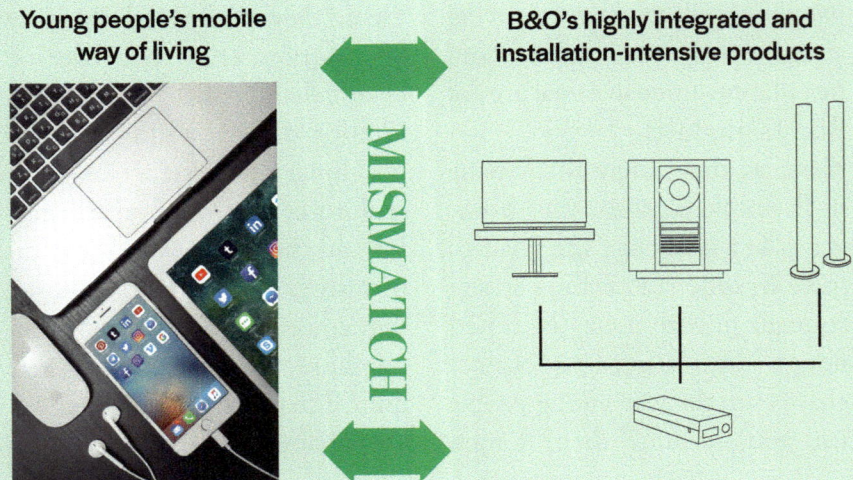

*Figure 8.3: Mismatch between young people's mobile way of living and B&O's
highly integrated and installation-intensive products*

**Sculptures of hi-fi
excellence**

**Young people's home
interior**

MISMATCH

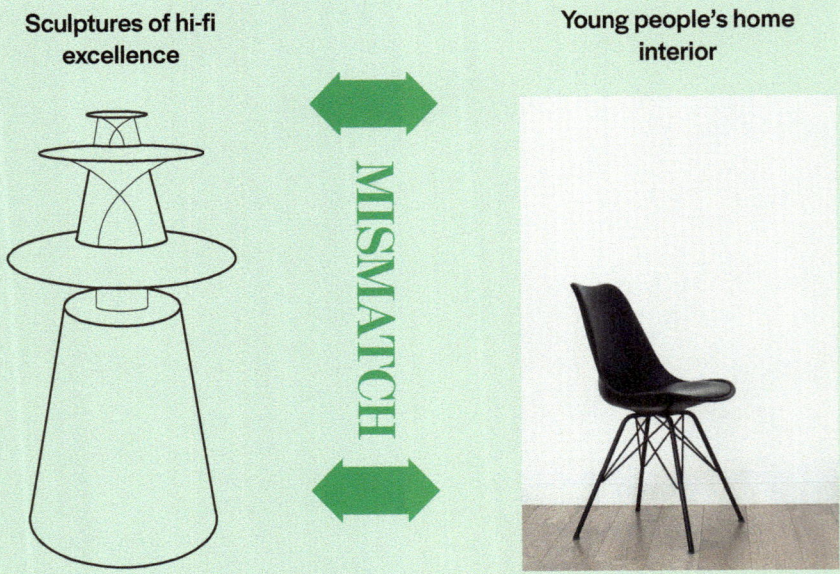

*Figure 8.4: Mismatch between sculptures of hi-fi excellence and the interior design
of young people's homes*

young families were more likely to invest their money in art or sculptures, which had for decades inspired B&O to make large sculptures of hi-fi excellence.

STEP 2:
Reframing the strategic fit with the younger audience

Having identified the mismatches between young people's aspirations and the products in B&O's current portfolio, the expert designers used the knowledge they had developed regarding the mismatches to reframe the new product's strategic fit with users, thereby creating a new design direction for the product. They determined that the new product needed to be more like a 'designer chair' and less like the 'sculptures of hi-fi excellence' the company had marketed previously. In particular, the well-known Eames chair that was experiencing a revival at the time served as a key inspiration for B&O's new visual design direction. Hence, the 'It is more like a designer chair than a pretentious sculpture' design principle became a key design principle behind the new product (Table 8.5).

In a similar way, the expert designers used the derived knowledge concerning the mismatch between the younger audience's mobile way of living

---

Design principle 1:

IT IS MORE LIKE A DESIGNER CHAIR THAN A PRETENTIOUS SCULPTURE

Reframing the strategic fit with the user:

There is a mismatch between the sculptures of hi-fi excellence and the interior design of young families' homes. The new product's identity should be more like that of a chair and less like that of a sculpture.

Move away from:	Be like:
B&O's sculptures of hi-fi excellence	A designer chair

Table 8.5: Design principle 1

Table 8.6: Design principle 2

and the installation-intensive nature of B&O's products to reframe the new product's strategic fit with users. The expert designers knew that the new product needed to move away from the idea of full home installation and towards a more mobile and flexible approach to music that would allow the speaker to be moved around. This new design principle was termed 'On the go' and is illustrated in Table 8.6.

The outcome of the expert designers' reframing was the B&O A9 speaker. The A9 is a round speaker with a fabric cover and wooden legs that give it the 'feel of a chair', which makes it easy to fit into a contemporary Scandinavian home interior. Moreover, it goes where the party people are. It is easily transportable due to the handle on the back and airplay connectivity, which makes it easy to move around the house, to the garden or from house to house. The only installation it requires is being connected to a power plug.

STEP 3:
Realigning the existing strategic fit

During the design process, the expert design team also made sure that the new product would have a lasting strategic fit with both the market and the company.

As the existing products in

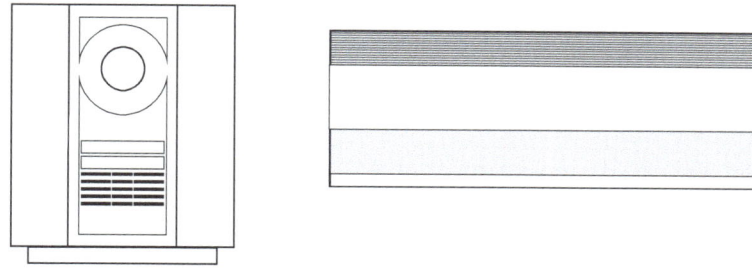

Figure 8.7: Beocenter 2300 and Beomaster 2300

the portfolio had been able to create lasting product-market and product-company fits, the expert designers identified the design principles behind these strategic fits and transferred them to the new product in a renewed form. They did so by investigating some of B&O's most significant and iconic products, that is, products that had been successful for a long time and products the company was particularly known for. In other words, they examined the core B&O products.

The expert designers studied these core products in order to identify how they created long-lasting strategic fits. They then identified the common design principles across the different

Design principle 3:

MAGICAL INTERACTIONS

Strategic fit with the market (differentiation):

Many core B&O products offer magical interactions that differentiate from other products and that exceed customers' expectations

Solutions in the core B&O products:	Solutions in the A9:
• Touch panels • Glass that slides open	• Turn music on and off, as well as turn the volume up and down, by touching the top of the product

Table 8.8: Design principle 3

'core products' that created lasting product-market fits or product-company fits.

The expert design team found that one design principle in particular had created a lasting strategic fit with the market. This principle was termed 'Magical interactions' (Table 8.8). Many core B&O products offered these magical interactions, which differentiated them from other products on the market. The intention behind the magical interactions was to create features that exceeded customers' expectations. Examples of these magical interactions can be seen in the Beocenter 2300, where the glass panels open when you move your hand close to them, and in the smooth touch panels of the Beomaster line, which contrast with the buttons that control the other products on the market (Figure 8.7).

In terms of the A9 speaker, 'Magic interactions' was a significant principle that was adopted in a renewed form by allowing the customer to control the speaker through merely touching it. During the design process, the expert designers had noticed that access to unlimited music services (e.g. Spotify) had made it difficult for users to select music and slowed down the process of turning on music when compared with regular radio. Their aim, therefore, was to create a product that would

Design principle 4:

B&O CRAFTMANSHIP

Strategic fit with the company:

All B&O products demonstrate the value of craftmanship.
It is central to both the brand and the company's values.

Solutions in the core B&O products:	**Solutions in the A9:**
• Precision in all assembly points • Expertise in aluminium surfaces	• Precision in all assembly points • Tight front cover

Table 8.9: Design principle 4

Table 8.10: Design principle 5

start playing easily and instantly. Thus, the A9 can be turned on by a single touch on its top (and it will play the exact radio station or playlist it was playing when it was last turned off). Similarly, the volume can be turned up and down by touching the top of the speaker.

While examining the core products, the expert designers also identified several principles that had created a strategic fit with the company. The first of these principles was termed 'B&O craftmanship' (Table 8.9). Many of B&O's iconic music products were known for their craftsmanship, which was central to the company's values. During the creation of the A9, craftsmanship also played a central role. Nothing was left to chance in terms of the detailing or assembly. In fact, during the development of the A9, the expert design team faced a number of challenges regarding the fabric cover on the front of the speaker. If they tightened it too hard, the fabric would wrinkle along the round edge. However, if the cover was not tightened enough it would move and flap slightly when the volume was turned up. Neither of these solutions lived up to the 'B&O craftmanship' design principle. In the end, the design team turned the bass round so that the low-frequency sound came out of the back through a decorative perforation. Moreover, the aluminium ring around the edge of the product reflected the B&O heritage, and the (re-) introduction of highly crafted wood gave a furniture reference linking back to Scandinavian design traditions.

Another principle that had created a lasting product-company fit was termed 'B&O geometric'. The most iconic B&O products were often shaped by simple geometric shapes that could be read as a silhouette. For the A9 speaker specifically, the expert team found the circle as a central visual element of musical instruments (tuba, trumpet, etc.) or music objects such as records, compact discs and the volume button. This visual principle was also transferred to the A9 in terms of its main shape. The principle is illustrated in Table 8.10.

Expert strategy 3 in practice

The B&O case is an interesting example of how insights into other product areas and product categories can reveal users' new and emerging needs, behaviours and expectations and, consequently, inform the direction of new product concepts. Who would have thought that changes in interior and furniture design could be so central to the reframing of a new speaker?

During our research, we have met a number of expert designers who are highly sensitive to such changes in users' needs, behaviours and expectations. These designers just need to look at a product for a short time to understand the kind of change it drives or captures. It is almost as if they always pick up on and collect new ways of living, behaving and interacting. This is an incredible capacity and competence, and it takes many years to become skilled at noticing these things and to collect enough material in one's personal mental library to be able to bring it to bear when necessary.

In the action guide for strategy 3 (see Chapter 13), we have established a process for identifying emerging changes in users' behaviours, needs and expectations. This includes all searches for changes as well as the identification of inspirational products that either drive or capture such changes. Finally, the action guide also includes guidelines for renewing lasting product-market and product-company fits.

Expert strategy 4: Extending product value
Expert example: Coloplast SenSura Mio
Expert strategy 4 in practice

EXPERT
STRATEGY 4

09

STRATEGIC FIT CHALLENGE 4

Strengthening the product-market fit

The fourth strategic challenge involves designing a new product for a portfolio where some previous products did not have strong fits with the market. More specifically, the challenge involves strengthening the product-market fit and ensuring that the new product will be competitive in the market in the long term.

EXPERT STRATEGY 4

Extending product value

The main focus of this strategy is on extending the value that the product provides to customers, for example, by providing new services or covering new emotional or social dimensions that allow the product to differentiate itself from other products on the market and, therefore, consolidate its strong position on the market.

Expert strategy 4: Extending product value

The fourth expert strategy is termed 'extending product value'. In the case of product development, the company's aim is to extend the value that the new product provides to customers. This strategy is particularly relevant for long-lasting market-leading products. Market-leading products provide the customer with more value than competitors' products; however, over time competitors typically improve their products and, at some point, are able to provide the same value. In other words, they catch up because they become able to offer the same level of quality, the same set of features or the degree of same performance as the market-leading company. When new products are developed, the main focus is on finding ways to extend the value that the new products will provide to users and customers so as to create lasting strategic fits. This can be done by, for instance, identifying the emotional or social values the product can provide to users and customers.

When working with this strategy, the expert designers follow three key steps: 1) identify ways to extend the value, 2) reframe the fit with the market and 3) realign the current strategic fit.

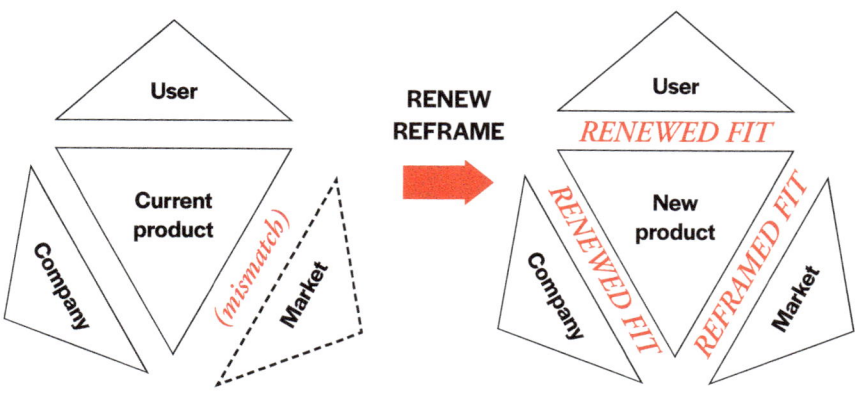

Figure 9.1: Expert strategy 4: Extending product value

Step 1: Identifying ways to extend the value

At the beginning of the design process, the expert designers identify the value that both the company's existing products and competitors' products provide to customers during their lifetimes. The expert designers map such value in order to identify areas where the customer experience can be improved as well as where functional, personal or social value can be added.

In particular, the expert designers focus on identifying areas where current products create unwanted experiences, interactions, emotions or social perceptions (e.g. things that make the user create workarounds, things that make the user feel incompetent, annoyed, irritated or uncomfortable or things that may create a negative perception of the user from the perspective of other people). Moreover, some expert designers also examine whether the products create value for an extended group of stakeholders and if there is the potential to extend the value for them in some way.

Step 2: Reframing the fit with the market

Next, the expert designers use the knowledge they have developed regarding the potential to extend the product's value as a means of reframing the product's strategic fit with the market. They do so by identifying inspirational products or services that create the type of value they are trying to achieve. These inspirational products are most often not from the same product category or user group. Rather, they are selected because they create a specific and lasting value.

The expert designers identify the design principles behind the inspirational products that are particularly central to the creation of the specific value. In some cases, the design team even samples design principles from many different types of products that create the same type of value that they are aiming for.

During the design process, the expert designers use the design principles derived from the inspirational products to reframe the new product's strategic fit with the market, thereby adding new value to the new product and differentiating it from the other products on the market. Sometimes, the expert designers even directly transfer design principles from the inspirational products to the new product design.

Step 3: Realigning the current strategic fit

Throughout the design process, the expert designers ensure that the new product will have a lasting strategic fit with the user and the company. They achieve this by reviewing the ways in which the products in the current portfolio create such fits, that is, products that have lasted for a long time, products that have been particularly successful or products that the company is particularly known for. As these products have proved to create long-lasting product-user and product-company fits, the expert designers identify the common design principles that have generated these strategic fits and then transfer them to the new product in a renewed form.

Expert example
COLOPLAST SENSURA MIO

STEP 1:
Identifying ways to extend the emotional value

The first thing that the expert design team at Coloplast did was to identify the value that both the company and its competitors provided to users and where there was potential to improve that value. In particular, they identified some potential with respect to the emotional challenges of undergoing an ostomy. For many people, undergoing an ostomy operation leaves them feeling as if they might as well have died. This feeling occurs because they are confronted with a number of taboos and trans-border experiences and because of the intense fear of never being able to live a normal life involving

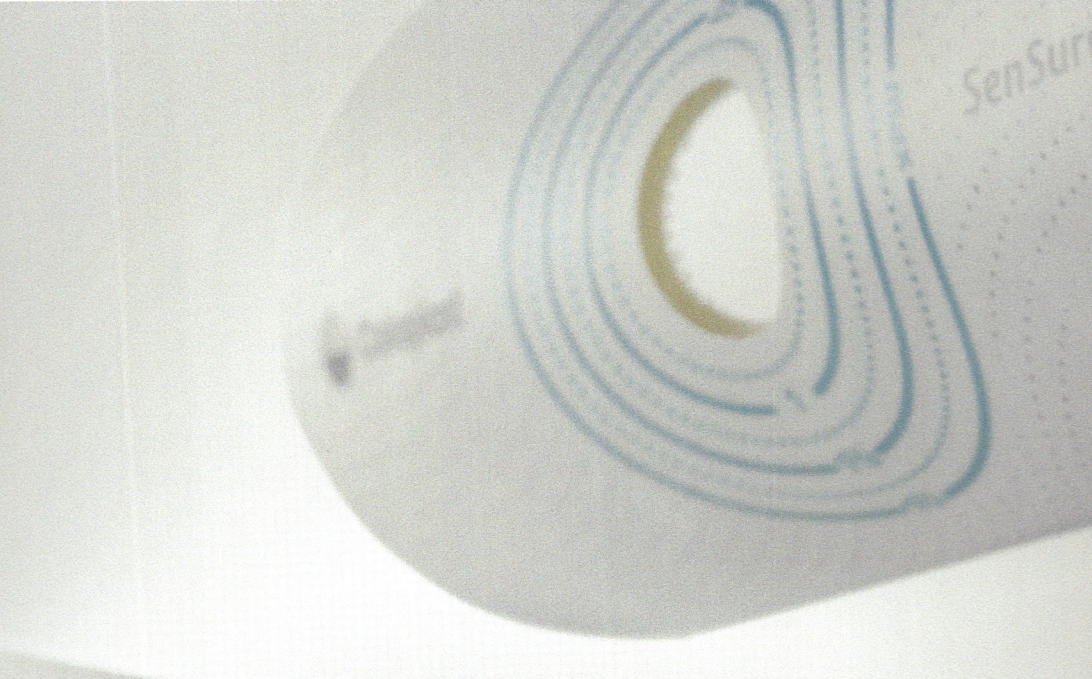

Figure 9.2: SenSura Mio. Credit: Coloplast

Table 9.3: Design principle 1

socialisation and intimacy again. However, after a year or so, things start to change and the emotional upheaval becomes less prevalent. At this point, all the new challenges associated with life after an ostomy start to fall into place and become part of everyday routines. That is, except for one thing. The ostomy bag still looks and feels like a medical product. Having to wear a medical product every day feels stigmatising, and it is an unwelcome reminder of being a patient or having a disability.

STEP 2:
Reframing the strategic fit with the market

Based on the knowledge gathered regarding the emotional challenges of the diagnosis as well as the insights into the medical look of the ostomy bag and feeling of being stigmatised, the expert designers started to think about products that would foster the opposite feeling. That is, the feeling of being normal. They thought of products that people wear every day, such as clothing, underwear and the like, and these products became the driver of the reframing: What should the new product feel like?

Wanting to relieve the user of the feeling of being sick, the designers came up with the idea of making an ostomy bag that looked like a piece of clothing customers would wear every day. This new perspective on the ostomy bag became the first design principle associated with the new product (see Table 9.3). The framing of the ostomy bag 'as a piece of clothing' highly

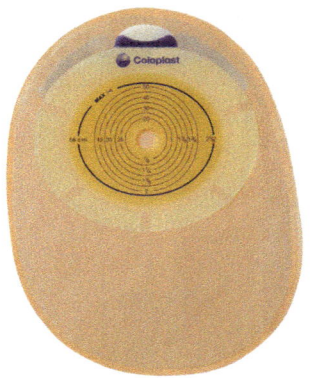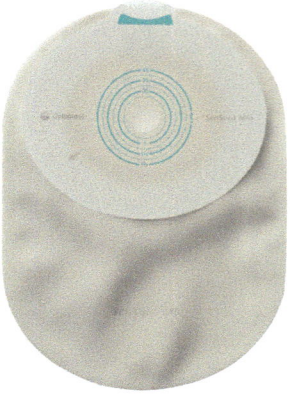

Figure 9.4: The old version of Coloplast's ostomy bag and SenSura Mio
(Credit: Coloplast)

influenced the shape of the new product. Unlike most ostomy bags, which were organic or 'kidney shaped', the new ostomy bag was created from geometrical shapes that resembled the cut and shape of most clothing.

Given this new framing of the product, the expert designers dug deeper into the materiality of clothing and other wearables. In particular, in the clothing category, the designers identified a 'material dignity' associated with clothing that contrasted with medical products. Products that are designed to be worn close to the body typically have a soft and comfortable sensation, and they are often made from woven materials. By contrast, ostomy bags are usually made from plastic materials typically used in medical products such

Design principle 2:
MATERIAL DIGNITY
Reframing the strategic fit with the market

Move away from:	Be like:
Plastics and medical materials	Woven material that feels comfortable to the skin

Table 9.5: Design principle 2

Table 9.6: Design principle 3

as disposable covers or aprons. Thus, 'material dignity' became the second design principle associated with the new product as well as the reason why the new product was made from woven fabric. Coloplast had not previously worked with woven fabric and introducing this stretchable material into its well-optimised production line caused some challenges. Nevertheless, the aim of bringing 'material dignity' to the product was eventually achieved (Table 9.5).

Looking at the inspirational products (i.e. everyday wearables), the expert designers identified another important thing, namely the ability to wear a white shirt without people being able to see what you are wearing underneath. For those who wear an ostomy bag, the aim here is to avoid people seeing the ostomy bag. During the investigation that followed, the expert team realised that it was not necessarily the bag itself that could be seen through a white shirt; rather, it the shadow it cast on the skin was visible. Hence, the third design principle behind the new product was 'To become one with the shadow'. This principle initiated an in-depth examination of which colours matched with the shadow of an ostomy bag. Ultimately, a light grey colouring was identified as being ideal for 'becoming one with the shadow' and, therefore, the new product was made in this colour. This change to a grey colour from the previous 'skin' colour also provided the product with a new and extra value for customers. Previously, one of the challenges of ostomy bags had been that even if the bags were 'skin' coloured, they would never fit the true colour of each user. Moreover, the use of 'skin' colour could even be stigmatising for people from different ethnicities. The new grey colour was neutral, and 'becoming one with the shadow' suddenly added a new extended value to the product (Table 9.6).

STEP 3:
Integrating the reframing with Coloplast's current strategic fit

During the design process, the expert design team at Coloplast ensured that the new product would have a lasting strategic fit with the user and the company. They did so by identifying the ways in which Coloplast's previous ostomy bags created such fits and then transferred them to the new product in a renewed form. Essentially, Coloplast's products had created a strong fit with the user because they were very safe to use. The first bag had been created with the aim of enabling people who had undergone an ostomy to re-establish their social life following the operation and, therefore, avoid isolation. Thus, a key aspect of any of Coloplast's products is that the user of the ostomy bag feels secure and that the bag will not leak any fluids or smells. Users must trust the product and feel secure when using it, which gave rise to the 'Security first' design principle (Table 9.7).

Design principle 4 (CORE):
SECURITY FIRST
Transferring current strategic fit with the user and the company

Aspiration:	Solution principles:
To make the users feel secure using the product, and never experiencing any leaks or smell exposure	• Secure adhesive disk that is gentle to the skin • Double layer system to prevent leakage

Table 9.7: Design principle 4

Expert strategy 4 in practice

The SenSura Mio case is a fine example of how it is possible to extend a product's value in order to create a lasting competitive advantage in the market. In our research, we have observed multiple examples of expert designers who are highly skilled at identifying social or personal needs that can add a whole new dimension to a product category. However, we must acknowledge that the ability to identify long-lasting needs, as well as the capacity to develop excellent product or service ideas that can result in a unique and long-lasting market position, is highly dependent on expertise.

In the action guide for strategy 4 (see Chapter 13), we suggest several tools that can help you to identify new and unmet emotional needs. Moreover, the action guide also reviews how to renew a lasting strategic fit with the user and the company.

EXPERT
STRATEGY 5

STRATEGIC FIT CHALLENGE 5

Framing a new and lasting strategic fit

The fifth strategic challenge involves creating from scratch a long-lasting strategic fit when it is not evident from the outset what to build on and what to avoid.

EXPERT STRATEGY 5

Searching for hooks

The main focus of this strategy is on creating from scratch a lasting strategic fit with the user, the market and the company.

Expert strategy 5: Searching for hooks

The fifth expert strategy is termed 'searching for hooks'. The main focus of this strategy is on creating from scratch a lasting strategic fit with the user, the market and the company.

Sometimes, companies that produce long-lasting products need to identify markets or market segments that are underserved so that they can market their product in new directions. A product's longevity means that its replacement will involve a limited and slow process. Therefore, the best strategy is to enter new markets with the product either in its current form or in an altered form. If there is a match between the market and the company's competencies, technologies or resources, it is identified as a market opportunity, and the design team's challenge involves creating a version of the product that targets the new market. As it is a new market and segment for the company, the strategic challenge involves generating long-lasting product-user, product-market and product-company fits from scratch.

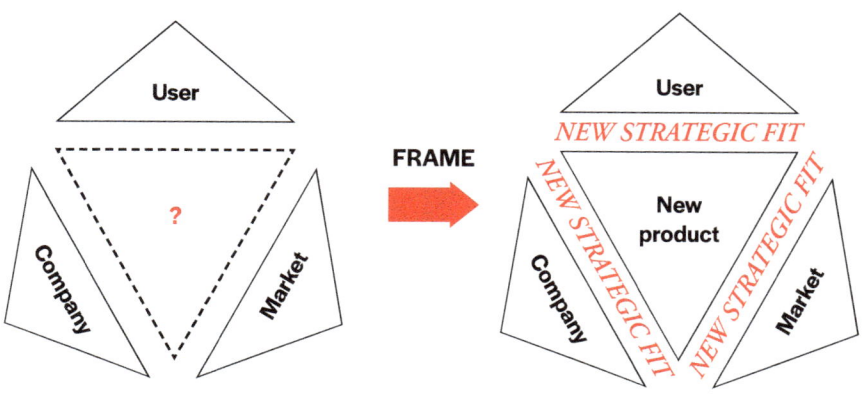

Figure 10.1: Expert strategy 5: Searching for hooks

One of the most significant pitfalls when creating a product for a new market is to mindlessly adopt beliefs and assumptions regarding the company's existing market and users and then use them as the basis for the new product (e.g. thinking that they already know what users want or what is attractive on the market). Creating a product for a new market means that there is nothing to reframe or renew—it is important to start from scratch and depart from previous beliefs, habits and assumptions. Knowledge concerning users and markets is deeply rooted in many companies, which means it can be unintentionally transferred to the creation of a new product if care is not taken.

Thus, the focus of the expert designers when implementing the fifth strategy is on continuously reminding themselves to start from scratch and create an ***entirely new and lasting*** strategic fit with the user, the market and the company. They do so through an interactive process of exploring, creating and testing.

When implementing the fifth strategy, the expert designers follow three key steps: 1) identify the 'ingredients' that hook the user's attention, resonate with them in the long run and frame the product-user fit, 2) identify the design principles that create a long-term competitive advantage on the market and frame the product-market fit and 3) continuously align with the company's values and core competencies in order to ensure a long-term strategic fit.

Step 1: Framing the product-user fit from scratch

The first step involves framing a lasting strategic fit with the new user group. Even if the users in that group seem quite similar to some of the company's other users, the expert design team acts as if they are starting from scratch. First, they search for design principles that will ensure that the product will be highly attractive to the users and resonate with them for a long time. As one of the expert designers explained: 'What are the key ingredients that the [users we target] really, really like?'.

During this process, key insights concerning the new users are collected. Moreover, the design team often identifies a set of inspirational products that have been particularly successful in terms of creating a lasting strategic fit with the users. It could almost be argued that the expert design team gets to know the new user group through the products that really resonate with them.

When analysing the inspirational products, the expert designers do not focus on product specifications or benchmarks for the new product to live up to. Instead, they search for the key design principles behind the inspirational products that ensure long-term strategic fits with users.

Next, the expert designers develop multiple product ideas or product concepts based on the key insights derived regarding the users and the design principles behind the inspirational products. These ideas and concepts are continuously tested with users in order to determine which ones 'hook' users' attention and which do not.

The overall process is highly iterative and involves several loops of creating, testing, recreating and testing until an idea or concept is finally found to create a promising long-lasting product-user fit.

Step 2: Framing the product-market fit from scratch

The second step for the expert designers involves ensuring that the new product holds a lasting competitive position on the market. Here, the expert designers are highly conscious of the need to not adopt any previous beliefs or assumptions stemming from the company's existing markets. Rather, they explore some of the most successful and long-lasting products currently on the new market, search for inspirational products that have remained competitive for a long time and identify the design principles that have created lasting product-market fits.

Next, the expert designers engage in an idea and concept development process whereby they apply the identified principles in different ways and continuously test the outcomes with customers, sales personnel, marketers, etc. to determine what works. Similar to the framing of the product-user fit, the expert designers engage in a highly iterative process that involves several loops of creating, testing, recreating and testing until an idea or concept finally proves to create a long-lasting product-market fit.

Step 3: Framing the product-company fit from scratch

The third step for the expert design team involves ensuring that the new product will have a lasting strategic fit with the company. The expert designers achieve this by continuously presenting the concepts and ideas to internal stakeholders in order to determine their likes and dislikes and, therefore, to become increasingly aware of which design principles create a strategic fit with the company and which do not. In many ways, the strategic fit with the company needs to be negotiated, and it is important that all objections are heard. When entering a new market, it is often not possible to renew or adopt a strategic fit from a core product. Indeed, in most cases, the new strategic fit with the company needs to be framed from scratch.

Expert example
LEGO NINJAGO

STEP 1:
Framing the product-user fit

In the Ninjago case, the designers were highly conscious of the need to avoid adopting beliefs and knowledge regarding LEGO's traditional users when designing the new product. They started the project with some small explorative tests so as to obtain initial insights into the new user group. Besides developing an initial understanding of how the action-oriented children differed from the LEGO Group's traditional construction-oriented users, the research focused on understanding what kind of 'worlds' (e.g. jungles, oceans, deserts, etc.) the new users were attracted to.

The exploration did not provide

Figure 10.2: LEGO Ninjago. Credit: iStock.com/abalcazar

any insights into which 'world' the action-oriented children were most attracted to. However, it caused the design team to realise just how important the story was in terms of igniting the play. As one of the designers explained: 'You could have a kid that really liked the desert. His favourite animals were a shark and a lion, and then he tried to come up with the story featuring a shark and a lion and the desert. It didn't resonate with any of the other kids in the room.'

After the design team had identified the story as being particularly important to the action-oriented children, they initiated a search for the key principles behind successful and long-lasting stories. In particular, they looked at some of the most successful stories from children's movies and determined that part of their huge success was down to the fact that they were based on themes most children know and are attracted to. 'Hey, all kids know or find the lion family really, really interesting! [...] They're similar to humans... with the hierarchy. They take care of each other and, of course, they also have their fights within and so forth.'

Hence, the designers concluded that an open and attractive starting point for the story (such as lions or toy characters) was highly important when seeking to create a lasting strategic fit with new users as well as an important design principle for the new product (Table 10.3). During the conceptualisation

---

Design principle 1:

OPEN AND ATTRACTIVE STORY

Framing the product-user fit

Inspiration:	Solution principles:
The most successful stories from children's movies	• A theme all children know and are attracted to • A theme all children can understand and engage with

Table 10.3: Design principle 1

of Ninjago, there were multiple loops of identifying key design principles, creating ideas and concepts, creating visualisations and models, and testing those creations with users until something appeared to resonate. These activities were initiated to develop different story elements, including the heroes of the story, their mission, their enemies, etc.

STEP 2:
Framing the product-market fit

The second step for the Ninjago team involved ensuring that the new product would also hold a lasting competitive position in the market. The key question was as follows: How do we create a story that is broad enough to be attractive to many different children and, at the same, is different and recognisable from all the other stories that children are introduced to?

By looking at some of the most successful and long-lasting stories within films, games, toys and the like, the design team realised that most of these stories had a sense of uniqueness because they had a 'tweak' to the story. This tweak provided a new perspective or new dimension that rendered the story more interesting (such as an anti-hero or mutated heroes in different other stories): 'I mean, in the toy industry you need to come up

---

Design principle 2:

A STORY WITH A TWEAK

Framing the product-market fit

Inspiration:	Solution principles:
Stories with a tweak (such as an anti-hero or mutated heroes)	• Uniqueness through combinations of different story worlds • Use of opposites such as new and ancient, strong and weak, serious and fun

Table 10.4: Design principle 2

with a novelty that is so new to the kids that they really want to play in this world and talk about it. If it's just knights... then it's not new enough. [...] Make that tweak so that it is really novel.'

The design team created ten different stories, all with a tweak. The stories were drawn up and presented to children in the target group. 'We showed [the children] ten different worlds during this process. So if you talk to a hundred kids and more than fifty are pointing to one "world", you kinda get that feeling.' Most of the children were drawn to the Ninja-themed story featuring East Asian culture and myths but with the tweak of modern cities, futuristic vehicles and technology. Accordingly, the second design principle that the team built on was "A story with a tweak" (Table 10.4).

STEP 3:
Framing the product-company fit

During the development of Ninjago, the design team presented the project to members of LEGO's management team at various points and, through these interactions, the challenge of creating a lasting product-company fit became apparent.

On multiple occasions it was argued that the Ninjago project's limited marketing budget as well as the LEGO Group's position as a toy manufacturer rather than a film producer meant that introducing a new and complicated story to the market could involve something of an uphill struggle.

At the same time, the Ninjago concept had been successfully tested with a large group of children. As one of the lead designers explained: 'We did this test. The facilitator, who had been with us for over 25 years, said that she'd never seen so much energy in the room and so much passion in the kids. She actually said: "You have something really unique and you should do something about it". [...] We told this story to our leadership team and said that we wanted to create a TV series.'

The design team pushed to upscale the Ninjago concept to what they termed a 'power theme' and, eventually, their efforts paid off. It was decided that Ninjago would include both a television series and a larger marketing budget.

Having convinced the LEGO management team of the po-

Design principle 3:

POWER THEME

Framing the product-company fit

Inspiration:	Solution principles:
TV series with long-lasting power-themes for children	• Film • Toys • Bring-to-school experience (craze)

Table 10.5: Design principle 3

tential of the Ninjago theme, the designers engaged in a new investigation into what it would take to create a successful power theme. They initiated this investigation by looking at some of the most long-lasting power themes for children. Here, they saw that, in addition to creating a television series, one of the key principles behind the generation of a long-lasting power theme was the creation of a take-to-school toy experience that could start a craze among children (such as Pokémon cards and Beyblade spinners). This key insight became the driver of the development of the Ninjago Spinjitzu (see Table 10.5).

Expert strategy 5 in practice

Although the example of LEGO Ninjago may seem straightforward based on the case description, our research revealed that framing a new strategic fit with the user, the market and the company requires a lot of effort and experience. Even expert design teams make a lot of mistakes and develop numerous ideas and concepts that do not hook users' attention in the way that they hope. For example, during the development of the Ninjago concept, the expert team designed a concept idea that did not provide any fit with users at all. As one member of the design team recalled: 'The kids didn't actually understand [the concept]. [...] they liked the world, but they didn't like the mission, and they didn't understand how to play it out.'

Moreover, we observed numerous examples where assumptions and beliefs regarding the market or the user were not questioned as well as other issues that hindered the design team's ability to start from scratch.

In the action guide for strategy 5 (see Chapter 13), we propose numerous actions that should prompt you to start from scratch and support your testing of ideas and concepts with users, customers, sales personnel, marketers, managers, etc.

Designing for product longevity
Longevity challenges and strategic durability
Application of the expert strategies

1

EXPERT
STRATEGIES
AND LONGEVITY
CHALLENGES

Designing for product longevity

The five expert strategies for creating strategically durable products and ensuring long-lasting strategic fits with the user, the market and the company all play a role in the transformation from a linear economy to a CE. If products becomes more strategically, emotionally and functionally durable, they will be able to resist obsolescence and the associated resource loops will be slowed.

In particular, the slowing of resource loops has been highlighted as critical to success of CE. As Korhonen et al. (2018) argue, the closing of resource loops is not enough. The circle also has to be slowed. Otherwise, the shift will only result in circular products that consume the same amount of energy and exert the same negative environmental impact as comparable non-circular products. Designing for product longevity ensures that the inner loops of the circle will be slowed down, which is key to the sustainable transition. Yet, a slowing of resource loops will only be possible if it is supported by radical changes in the way products are consumed, developed and designed as well as by changes in the way business is done. New types of circular business models and value propositions are essential (Bocken et al. 2016; Haffmans et al. 2018), while new ways of handling the design process are required (Bakker et al. 2019).

Longevity challenges and strategic durability

This book contributes new knowledge to the literature concerning how design teams can address long-term user needs and wishes, create a long-term competitive advantage in the market and ensure that products will be relevant to the company in the long term.

It presents a detailed analysis of how expert design teams create strategically durable products with lasting strategic fits. We have reviewed how expert design teams focus on solving user needs that will be just as relevant in the future as they are now, for example, developing an ostomy bag that does not make the user feel stigmatised. We have also reviewed how expert designers create unique competitive advantages that render products attractive in the market in the long term, for example, by offering young girls a toy based on deep visions of friendship and equality in a girls' toy market otherwise dominated by the 'one girl in front' attitude. Finally, we have shared insights into how expert design teams create products that are able to reinterpret the visions and values of companies in ways that render the products almost iconic, for instance, by showing how deeply rooted values of 'craftmanship' and 'magical interactions' can be renewed and reinterpreted in a new speaker.

The main contributions of this book are the conceptual framework for strategic durability as well as the five strategies for creating long-lasting strategic fits (see Figure 11.1).

The overarching aim of this book has been to supplement current design strategies for product longevity. While these current design strategies provide valuable support for the later stages of the design process and for detailed product design, they only provide very limited insights into how product longevity should be handled during the strategic and conceptual phases of the design process, which is when many defining decisions concerning the product are made. The research presented in this book reveals that focusing on the strategic durability of the product is essential throughout the conceptual design process.

Strategic durability

Creating products with strategic durability entails creating products with long-lasting strategic fits. That is, products that address users' long-term problems and needs, create long-term competitive advantages on the market, advance the long-term credibility of the company, enhance the strategic strengths of the company and align with the company's values, purpose and culture.

Five expert strategies

Expert strategy 1: Renewing core principles
Expert strategy 2: Leveraging objections
Expert strategy 3: Foreseeing future mismatches
Expert strategy 4: Extending product value
Expert strategy 5: Searching for hooks

Figure 11.1: Overview of the book's contributions

Application of the expert strategies

Designing for strategic durability is just one of many new strategies that need to be implemented in order to ensure product longevity. Hence, the expert strategies set out in this book are by no means intended to stand alone. Indeed, they must be used alongside other strategies for resisting, postponing and reversing obsolescence (see Figure 1.5), including design strategies for physical and emotional durability, maintenance, repair, upgradability, refurbishment, remanufacture, etc.

Similarly, the expert strategies cannot simply be applied on their own as a tool or method. They must be supplemented by a paradigm shift—from a linear logic to a circular logic—that highlights the need for continuous effort in terms of slowing and closing resource loops. In the conceptual design process, this also involves a radical change in the way strategic fits with the user, the market and the company are created. The expert strategies are examples of this new logic, which can be summarised in the following ways.

New types of product-user fits

In the linear economy, the creation of a product-user fit is driven by new wants and new trends in behaviours, or even by a proposal that could lead to new types of needs and aspirations, which leads to faster product obsolescence and product replacement. In the CE, creating a fit with the user involves identifying long-term needs, wants and aspirations, and the design team must identify not only what works now, but also what will be relevant in the future.

New types of product-market fits

Likewise, the creation of a product-market fit in the linear economy focuses on outdoing competitors with new competitive parameters that align with current trends and tendencies. This will strengthen the company's competitive positioning, although it will also drive product replacement because there will be a constant battle as to who has the latest, greatest model. In a circular paradigm, the focus shifts towards the creation of long-term competitive advantages and credibility.

New types of product-company fits

Finally, in the linear economy the fit between the product and the company is commonly created by exploiting the company's present competencies and strategic strengths. However, in the CE the focus is also on enhancing the strategic strengths of the company and aligning with the company's values, purpose and culture, which all entail more long-term perspectives.

In summary, there appears no doubt that a shift from a linear paradigm to a circular paradigm is currently on the public and academic agenda. A useful first step in this transition towards more circular production and consumption patterns is the implementation in the conceptual design process of the five expert strategies for achieving strategic durability. In Chapter 12, we will discuss the road ahead for both practitioners and researchers.

The road ahead
For practitioners
For researchers

2

THE ROAD AHEAD

The road ahead

For the last five years, we have dedicated most of our time and effort to examining how products with strategic durability are conceptualised as well as how they come to life. It has been an extraordinary journey and we sincerely hope that we have been able to share enough insights for you to embark on a similar journey. We had two key aims when writing this book. On the one hand, we sought to contribute to the research on product longevity by introducing a new research agenda that focuses on the earliest and most strategic part of conceptual design process and by sharing the key findings of our work with expert design teams. On the other hand, because design is such an applied profession and because we wish to create an impact, we sought to write a book that was relevant and usable for practising design and development teams who are developing new strategic product concepts and striving to design products that last. It is our belief that the road ahead follows two diverging paths, one for practitioners and one for researchers.

For practitioners

There is no doubt in our minds that the ability to create **strategically durable products** will be an in-demand skill on the part of any designer, engineer or product developer in the future, and we hope that this book has provided a first step towards developing this ability. We acknowledge that it might not be easy for you as a practitioner to internalise the experts' sensemaking and strategies all at once. However, the next time you engage in the creation of a new project, we hope that you will keep the following key insights in mind:

- Different assignments pose different strategic challenges and, therefore, require different strategies.
- For a product to be long-lasting, it needs to have a lasting strategic fit with the user, the market and the company.
- Existing products in the portfolio hold important strategic information. If you analyse them, you will gain important insights into where the strategic fit is weak and where it is lasting, which will support your actions during the design process.
- A company's core products typically offer unique insights into lasting strategic fits that have the potential to be renewed.
- A company's less appreciated products may provide valuable insights into what needs to be reframed in order to transform a weak strategic fit into a new and lasting fit.
- Inspirational products from outside the company reveal important knowledge regarding how to create a lasting strategic fit, for example, with a specific group of users or with a specific market. Take advantage of this knowledge.
- Looking for hooks and objections is vital. Make sure to talk to all the people involved with the product, as their insights will be essential to the creation of a successful and lasting strategic fit.
- It is imperative to create, test, recreate and test until the new strategic fit is in place.

Will the five strategies make you an expert?

Reading about experts and expert strategies will likely raise the following question: Will the expert strategies make me an expert? We wish the answer was yes, but it probably is not. Based on our experience, successfully creating long-lasting strategic fits demands both in-depth contextual knowledge and long-term experience. However, we have some hints for you on how to extend the use of the strategies and build your expertise with respect to creating strategic fits.

First, one of the things we found the expert designers to have in common is the fact that they collect successful and long-lasting strategic fits all the time. They just can't help it. They have *a mental library* of different design principles that will create lasting strategic fits and that they can bring to bear when needed. They collect these principles from projects, from conversations, from product releases, from interviews—from everywhere—because the subject interests them and because it is the key to their work. Through understanding how different kinds of products create strategic fits with different user groups, different markets and different companies, the expert designers have a better starting point when seeking to create strategic fits themselves. This is also evident when you work with an expert designer. They appear to bring these unique and brilliant perspectives to a project without having done any research (although the truth is that they probably never stop doing research).

Second, it is very clear from our conversations with the expert design teams that they pay close attention to people's reactions (which we term hooks and objections). They look for honest reactions and they create prototypes—not for the sake of the product, but to provoke honest reactions and investigate what *really* resonates and what does not.

Finally, we have found that resilience with regard to failing is imperative. Failure is a significant part of creating lasting strategic fits. Some of the expert designers showed us projects that were total disasters, while others elaborated on the mistakes they made in past projects. Failure is part of practising the art of creating long-lasting strategic fits, and practising is the only way to build expertise. What appears to set the expert design teams apart from younger teams is their determination to move forward, rather than to linger over the analysis of what went wrong.

How to use the expert strategies in practice

The use of the five expert strategies will differ depending on the level of expertise within the team and your experience as an individual designer.

If you are a member of a young design team: We propose that you practise the strategies—try out the action guides, become familiar with the different approaches and integrate them into your practice. You should seek to renew the core principles, leverage objections, foresee future mismatches, extend product value, search for hooks and build up your mental library.

If you are an experienced designer: We suggest that you use the expert strategies to reflect upon your own practice. As designers, we have a tendency to apply the same strategy all the time. We use the strategy that best reflects our preferences, suits our comfort zone or, perhaps, has proved successful in the past. Knowledge concerning the different strategies can help us to apply the strategy that is most suitable for the task, rather than the strategy we feel more confrontable with. Hence, this book can be used to extend your own or your team's repertoire.

For researchers

The road ahead research-wise also offers the potential to extend your repertoire. With this book, we propose a new research agenda for designing for product longevity, one that focuses on the strategic aspects of the product and its conceptual development. The concept of *strategic durability* is, we hope, just the first of many models, strategies and approaches for addressing product longevity during the very early stages of product development. From our perspective, it is essential that we make products that address users' long-term problems and needs, create long-term competitive advantages in the market and are of strategical importance to the company. To achieve this, new methodologies, strategies and approaches are required. In addition, there are also some issues that have not yet been adequately investigated and that provide opportunities for further research.

In this book, we have not touched upon the issues of technology and digitalisation or on how they might influence the creation of a strategic fit or disrupt a product's strategic durability at some point in the future. This represents an important avenue for future research.

From a broader perspective, the organisational setup, knowledge management and development processes leading to the creation of strategically durable product concepts with long-lasting strategic fits have also not been addressed in this book. Yet, our conversations with the expert design teams have revealed that these matters are imperative when it comes to successfully creating long-lasting strategic fits.

In the same way, we propose to add product longevity to the research agenda concerning strategic design. To date, research into the strategic aspects of design has failed to address product longevity and the slowing of resource loops. Instead, the focus of the strategic design research has primarily been on the strategic value of design (e.g. de Mozota 1998; Lockwood and Walton 2008; Brown 2009; Martin 2009; Verganti 2009), which has contributed to the creation of new strategic design methodologies intended to advance corporate strategy (e.g. Holston 2011; Buijs 2012; Curedale 2013; Kumar

2014; Calabretta et al. 2016; Nixon 2016). The addition of the concepts of strategically durability to this field may also prove to be an important avenue for future research.

Last but certainly not least, we hope that our book will help reposition 'product longevity' on the sustainable research agenda and inspire researchers, practitioners and policymakers to sustain their focus on the inner loops of the CE model.

Applying the expert strategies in practice
How to use the action guides

L3

ACTION GUIDES

Applying the expert strategies in practice

In this book, we have shared our insights into expert teams' strategies for creating strategically durable products, including how they handle the many aspects, challenges and complexities associated with creating long-lasting strategic fits with the user, the company and the market.

Perhaps you are facing similar challenges in your company or design team? We hope that this book has inspired you to take on the challenge of creating strategically durable solutions that provide long-term value to your customers and generate viable long-term business for your company.

We acknowledge that considering all of the expert strategies at once might seem overwhelming at first and that applying the strategies in your practice might not be an easy task. However, this chapter will do its very best to help you with the process.

We have used our knowledge of tools, methods and approaches, as well as our expertise in teaching design and design thinking, to translate the expert designers' strategies into a set of step-by-step suggestions for how the five expert strategies can be used in practice, depending on your design situation.

How to use the action guides

First, we kindly remind you that every design situation is unique and that design processes must address this uniqueness. Thus, it is important to remember that the proposed strategies cannot replace the reflective and solution-oriented design process; rather, they are intended to complement the design process. Likewise, methods and approaches such as user research, ideation, prototype development, user testing, context safaris, portfolio review and competitive and market analysis remain relevant and very much needed.

If you already have solid design experience and a favourite process that you use, please continue to use it. If you do not have a favourite process yet, we recommend that you—aside from this book—find inspiration in the many good books that focus on the design process. Moreover, we recommend that you familiarise yourself with the work of the authors we have referenced throughout this book.

Once you have the relevant knowledge in place, the proposed strategies can be used to their full potential. The strategies will help you to prioritise your focus on the design process and support your sense- and decision-making.

Every action guide in this chapter refers to a previous chapter that describes the strategy in more detail and includes an expert example. This means that if you need more detailed context or additional information, you should go back and review the specific chapter to recap the strategy.

For each action guide, we have created some general worksheets that you can use freely in your design process. We recommend that you start to use them as early as possible during the conceptual phases, although it might also be relevant to continuously revisit and revise them.

***Finally, good luck with the creation of your
strategically durable product!***

ACTION GUIDE 1:
What is your strategic fit challenge?

The first step towards creating a strategically durable product or product line involves analysing the type of strategic fit challenge you and your design team are facing. It may take some time to fully characterise your strategic challenge, but here are some helpful suggestions:

1. Draw up your strategic fit challenge

To identify your strategic fit challenge, it might prove helpful to map out what you currently know about the strategic situation. The aim here is to identify where the current strategic fit is strong and lasting and where it needs to be strengthened. This analysis could be based on just a few existing products, although it is often a good idea to look at the whole product line or product portfolio. We have created a strategic challenge worksheet to support you in this (see worksheet 1).

2. Compare your challenge to the case descriptions

Next, we advise you to look closely at both the five strategic fit challenges and the five case descriptions presented in Chapter 4. By reading through the cases, you may be able to identify similarities between them and your own project, which may help you to identify the challenge. Moreover, even if the cases are not similar to yours, they may change your perception or further nuance your understanding of the challenge you are facing.

3. Find the matching strategy

After you have identified a strategic fit challenge similar to your specific situation, the next step involves applying the matching expert design strategy. In the following action guides, we advise you on how to apply each strategy in practice.

Product-user fit

Q1: Does the product resonate with
its users?

Q2: Does the product solve
long-term problems, needs, wishes
and aspirations?

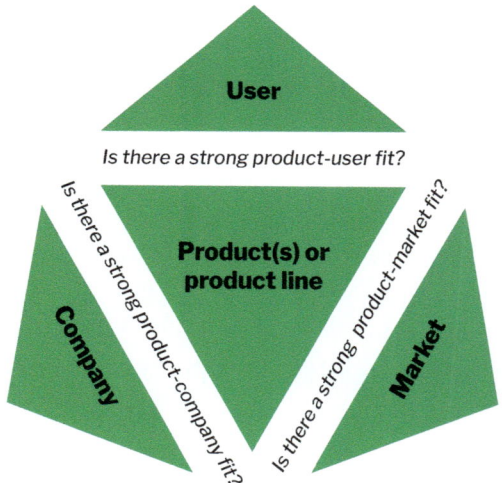

Product-company fit

Q3: Does the product build on
the company's strategic strengths?

Q4: Does the product take
advantage of the company's core
competencies?

Q5: Is the product aligned with the
company's core purpose?

Q6: Is the product aligned with the
company's core values?

Product-market fit

Q7: Does the product advance
credibility of the company and the
brand's position?

Q8: Is the company's competitive
position underlined or strengthened
by the product?

Q9: Does the product create
long-term competitive advantages
for the company?

Q10: Is the product unique and
differentiating in the market?

ACTION GUIDE 2:
Expert strategy 1

If you are creating a new product concept that is intended to assist already successful products in the portfolio and, therefore, to maintain the credibility of the company, it may prove useful to apply expert strategy 1 (as described in Chapter 6). This action guide provides some guidelines and supporting questions to help you transfer an existing long-lasting fit to a new product.

Step 1: What are the existing product's core principles?

We suggest that the first step you take involves identifying the core principles associated with the existing products that have created lasting fits with the user, the market and the company.

Initial core principle analysis

One way to initiate this analysis is to gather all possible material regarding the existing products, including marketing material, articles (newspaper/online) and product reviews, and then begin an initial identification of the existing products' core principles. Figure 13.1 presents some questions that can support this analysis, while worksheet 2 provides a template with which the core principles can be documented.

During this analysis, be careful not to just identify specific features of existing products that seem to resonate with the user, market or company. Instead, try to identify the principles behind those features. Core principles are generally supported by a number of different features and aspects. For example, in one of the cases in this book, the expert design team found that it was the 'professional tool-like feeling' of the products that really resonated with users. This 'professional tool-like feeling' was created by a combination of material choices, detailing, haptic feedback, etc. Together, these different elements created the feeling of being professional when the user used the product. Thus, when analysing the core principles, if you discover, for instance, the haptic feedback of existing products to be highly appreciated by users, try to determine why that is and what value it provides.

What are the design principles behind the existing products that solve long-lasting problems for users and address users' long-term needs, wishes and aspirations?

How do the existing products' design principles leverage the company's core competencies and long-term strategic strengths?

How do the existing products' design principles create long-term competitive advantages for the company and render the products unique and difficult for competitors to copy or imitate?

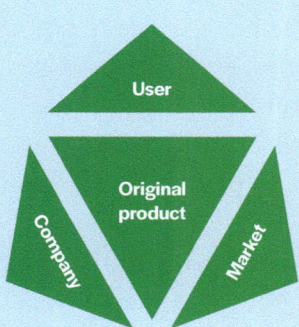

What are the design principles that render the existing products attractive to the company given its long-term values, purpose and culture?

How do the existing products' design principles advance the long-term credibility of the company and, therefore, its competitive position in the market?

Figure 13.1: Model for analysing the design principles behind existing products

Interviews with core stakeholders

In the quest to identify core principles, it is also a good idea to interview the designers or developers behind the existing products or the marketers, sales personnel, etc., who have played a significant role in the product's introduction to the market. Interviewing people who have been with the company for a long time is preferable. Based on these interviews with core stakeholders, it should be possible to identify the core design principles behind the existing products, which can otherwise be very difficult to discover. Moreover, the interviews may add nuance to some of the core principles behind existing products that you have already discovered. Worksheet 3 includes an overview of the types of questions that it can be relevant to ask core stakeholders. However, be aware that a bespoke interview guide needs to be developed for the interviews.

Step 2: How can you renew the core principles?

The second step in this strategy involves transferring the core principles that created the successful products' strategic fits to the new product. Unlike step 1, this does not entail a specific analysis accompanied by a specific set of approaches. Instead, step 2 must be implemented continuously throughout the conceptualisation of the new product. The transfer and renewal of the core principles must form part of the ongoing process of developing ideas, concepts and prototypes.

In accordance with the approach to reflective practice described by Schön (1983), we suggest using the knowledge concerning the core principle as knowledge in action. This means that the knowledge regarding the core principles should sit at the back of your mind while you develop ideas, concepts and prototypes. You should focus your attention on the concept or prototype you are creating and apply the knowledge regarding the core principle when doing so makes sense. This process is known as reflection in action.

Then, once you have complete the model or prototype, it is time to evaluate if the core principle can play a more central role. For example, if you have created several concepts and you want to determine which one to move forward with, you should bring together a list of core principles and then reflect on the extent to which the different concepts implement and renew the core principles. During this reflection, it may be possible to come up with ways to renew some of the core principles or to combine different concepts in order to allow for them to be implemented in a stronger fashion.

As with the analysis, you should be very careful during the conceptual process not to just copy features from existing products. Rather, you should focus on ensuring that all decisions and features concerning the new product are increasingly aligned with the core principles.

Step 3: How can you strengthen the strategic fit?

During the process of creating the new product concept, it is important to remember that the strategy is intended to complement the design process, not to replace it. The task is still to create a new product concept that will generate a long-lasting strategic fit with the user, the market and the company. Given that you should complement an existing and successful product with a new one, there are certain core principles that you can leverage from when you were creating the strategic fit, although doing so remains a complicated challenge. Therefore, it is of utmost importance to perform user research, context safaris, competitive and market analysis, portfolio review and analyses of the company's strategic strengths to identify areas where the long-lasting strategic fit of the new product can be either strengthened or renewed.

As is the case for the expert designers, it is relevant to search for significant changes in the market, changes in customer expectations and/or changes in the competitive landscape that need to be addressed by the new long-lasting product.

Product-company fit

Design principle 1:

Strategic fit with the company

Solution elements of the existing product

Design principle 2:

Strategic fit with the company

Solution elements of the existing product

Design principle 3:

Strategic fit with the company

Solution elements of the existing product

Product-user fit

Design principle 4:

Strategic fit with the user

Solution elements of the existing product

Design principle 5:

Strategic fit with the user

Solution elements of the existing product

Design principle 6:

Strategic fit with the user

Solution elements of the existing product

Product-market fit

Design principle 7:

Strategic fit with the market

Solution elements of the existing product

Design principle 8:

Strategic fit with the market

Solution elements of the existing product

Design principle 9:

Strategic fit with the market

Solution elements of the existing product

Aim of the interview

The aim of the interview with the core stakeholders is to get deep insights into the product and the reasoning behind the product. In order to achieve this, it is a good idea to start the interview by first going through the design process of the original product. This makes it possible to ask more direct and specific questions about the product, its principles and the reasoning behind it, later in the interview.

Moreover, it is unlikely that the core stakeholders can articulate the original product's core principles directly. The core principles are more likely central elements of the narratives they tell. Once the interview is completed, it can therefore be helpful to listen to it a number of times, in order to identify the core principles. Search for the key metaphor or one-liner (e.g 'larger than expected') and then find their connection to the strategic fit and the specific features that materialise the principle.

General questions that need to be contextualised in the interview:

The process:

What was the background/driver or initial idea behind the product?

What was the initial vision for the product?

How did the product develop over time?

What was the key insight (market, users, technology etc.) that the product is based upon?

What were the most important activities during the design process? And which important insights/findings did you get from these?

The product:

Who was the main user of the product?

Which problems, needs, challenges were the product intended to satisfy?

Why is the product experience/expression and interaction the way it is?

Is there any new technology or new use of technology in the product?

How did the product differentiate from other products on the market, when it was introduced?

How was the product positioned compared to its competitors?

Which of the company's core competencies and strategic strengths did the product built upon?

Why did the company management find the product attractive?

The reasoning:

What are the key principles of the product? Where did these come from?

Were there any disagreements with respect to specific parts of the product?

What kind of things could the product never do? And why?

Were there any features or aspects that were deselected in the product? And why?

What kinds of changes were made to the product from concept to production? And why?

Were there any conflicting insights or information you had to deal with?

ACTION GUIDE 3:
Expert strategy 2

If you are creating a new product for a portfolio featuring many products with weak product-company fits, it may be useful to apply expert strategy 2 (as described in Chapter 7). This action guide provides some guidelines and supporting questions intended to help you to identify weaknesses in the current fits and determine how to strengthen the product-company fit with a new and lasting solution.

Step 1: How do you identify objections and strategic mismatches?

The first step that we recommend involves identifying all the potential objections to the unfitting products. These objections can help you to identify the underlying reasons for the products' strategic mismatch with the company. Figure 13.2 presents some guiding questions that may support you in this exploration.

Explore objections within the company as well as product-company mismatches

We suggest that you begin this strategy by exploring what made the unfitting products less attractive to the company. This may prove a bit tricky and be very difficult to identify in an interview, although you should try to identify the unfitting products' harshest critics within the company and then discuss the products with them. Another option is to pay attention to any objections to the unfitting products that may have been raised and to ask people to elaborate on their opinions. These informal critiques can provide valuable insights into the features, aspects and design principles of the products that are less attractive to the company. Moreover, you should compare the unfitting products to some of the company's most successful products. The aim is to identify how the successful products created a lasting strategic fit with the company, for example, by identifying how they

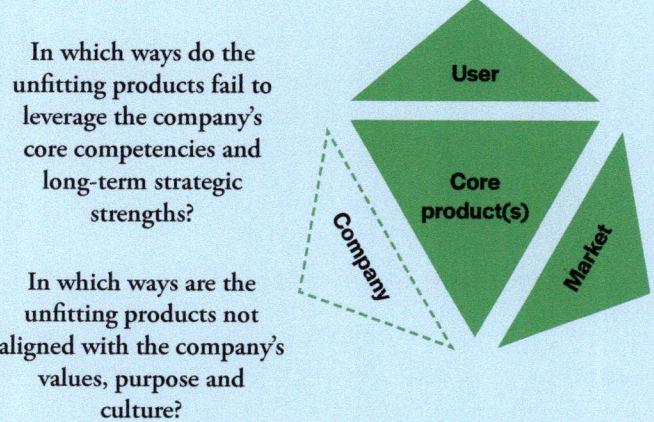

In which ways do the unfitting products fail to leverage the company's core competencies and long-term strategic strengths?

In which ways are the unfitting products not aligned with the company's values, purpose and culture?

Figure 13.2: Model for analysing the design principles behind existing products

integrate the company's strategic strengths and core competencies or the company's visions and values and then using this information to determine what the unfitting products might be missing. Worksheet 4 provides some guiding questions for interviews concerning the company's successful and core products, while worksheet 5 provides a template for gathering the key design principles that create either a fit or a mismatch with the company.

Explore credibility-related objections and product-market mismatches

In some cases, it might also be relevant to explore why the unfitting products are not considered credible in the market. To achieve this, it might be relevant to interview sales managers, marketers and other parties who are actively engaged in marketing or selling the unfitting products and to identify their objections to them. The aim is to understand the underlying design principles behind the products that somehow lead to them having a lack of credibility in the market.

Another way to do this involves comparing the unfitting products with some of the most successful and long-lasting products on the market. Through identifying the successful products' competitive advantages (or, to be more precise, identifying the 'design principles' that give these products long-term competitive advantages or lasting market positions), it is often possible to identify what the unfitting products might be missing and, therefore, the reason for their lack of credibility. In worksheet 6, you will find a guide for interviewing sales managers, marketers, etc.

Step 2: How do you reframe the strategic fit?

The second step in this strategy involves using the knowledge concerning the unfitting products' weak fit with the company as the starting point for reframing the new product's lasting strategic fit. This can be achieved by focusing on the reasons behind any mismatches and then suggesting ways to overcome them.

The reframing of the new product's strategic fit with both the market and the company has to form part of the ongoing process of developing ideas, concepts and prototypes.

In accordance with the reflective practice described by Schön (1983), we suggest using the knowledge concerning the design principles that created the mismatches and the 'design principles' behind the successful products that created a lasting strategic fit as knowledge in action. This means that the knowledge regarding the key principles should sit in the back of your mind when you are developing ideas, concepts and prototypes. You should focus your attention on the concept or prototype you are creating and then apply the knowledge concerning these design principles when it makes sense to do so. This is known as reflection in action.

Once in a while, it may prove relevant to let the identified design principles play a more central role. For instance, if you have created a number of concepts and you want to determine which one to move forward with, you should bring together a list of the fitting and misfitting design principles (e.g. from worksheet 5) and then reflect on the extent to which the different concepts serve to implement and reframe the design principles. During this reflection, it may be possible to come up with ways to combine different concepts in order to ensure they are more strongly implemented.

As with the analysis, you should be very careful during the conceptual process not to just copy features from successful products. We suggest that you focus on ensuring that all decisions and features related to the new product are increasingly aligned with the design principles that can create a long-lasting strategic fit.

Step 3: How do you test and align the strategic fit?

During the process of creating the new product concept, it is also highly important that you align the strategic fit in as many ways as possible. To achieve this, you should explore ways to make the product concept resonate with the user from a long-term perspective, ways to render the product competitive in the market in the long term and ways to render the product attractive to the company and make it align with its long-term visions. Moreover, as you are replacing an unsuccessful strategic fit, it is also important to continuously test the strength of the new product concept's strategic fit to see if it is lasting. This can be achieved by showing product ideas and concepts, including the earliest ideas, to users, sales personnel, strategic management and other key stakeholders throughout the development process in order to obtain as rich information as possible.

Aim of the interview

The aim of the interview with the core stakeholders is to get deep insights into the company's products and the reasoning behind them. In order to achieve this, it is a good idea to start the interview by going through the design process of selected products individually. This makes it possible to ask more direct and specific questions about the products, their principles and the reasoning behind them during in the interview.

Moreover, it is unlikely that the core stakeholders can articulate the design principles of the core products directly. The design principles are more likely central elements of the narratives, they tell. Once the interview is completed, it can therefore be helpful to listen to it a number of times, in order to identify the design principles. Search for the key metaphor or one-liner (e.g 'Pride of creation') and then find their connection to the strategic fit and the specific features that materialise the principle.

General questions that need to be contextualised in the interview:

The process:

What was the background/driver or initial idea behind the product?

What was the initial vision with the product?

How did the product develop over time?

What was the key insight (market, users, technology etc.) that the product was based on?

What were the most important activities during the design process? And which important insights/findings did you get from these?

The product:

Who was the main user of the product?

Which problems, needs, challenges were the product intended to satisfy?

Why is the product experience/expression and interaction the way it is?

Is there any new technology or new use of technology in the product?

How did the product differentiate from other products on the market, when it was introduced?

How was the product positioned compared to its competitors?

Which of the company's core competencies and strategic strengths did the product built upon?

Why did the company management find the product attractive?

How is the product aligned with the values and culture of the company?

The reasoning:

What are the key principles of the product? Where did these come from?

Were there any disagreements with respect to specific parts of the product?

What kinds of things could the product never do? And why?

Were there any features or aspects that were deselected in the product? And why?

What kinds of changes were made to the product from concept to production? And why?

Were there any conflicting insights or information you had to deal with?

Product-company fit
Design principles of the *inspirational* products

Design principle 1:

Strategic fit with the company

Solution elements of successful products

Design principle 2:

Strategic fit with the company

Solution elements of successful products

Design principle 3:

Strategic fit with the company

Solution elements of successful products

Product-company mismatch
Design principles of the *unfitting* products

Design principle 4:

Strategic mismatch with the company

Solution elements of unfitting products

Design principle 5:

Strategic mismatch with the company

Solution elements of unfitting products

Design principle 6:

Strategic mismatch with the company

Solution elements of unfitting products

Aim of the interview

The aim of the interview is to explore why the unfitting product is not credible on the market. Here it can be relevant to interview sales managers and marketeers or others, who are actively engaged in promoting or selling the unfitting product and identify their objections against it.

Voicing criticism might be a delicate matter inside an organisation, therefore it is highly important to angle the questions in the interview in different ways. In some cases, the interviewee might not even be aware of why the product is not credible on the market and therefore it is essential to give the interviewee time for reflection.

To ensure that the interviewee's opinions and objections are as precise as possible, make sure the unfitting product is available and discussed during the interview.

Moreover, ask the interviewee to bring/identify the products they find to be the most successful products on the market, and ask them beforehand to think about why they are successful compared to the unfitting product.

General questions that need to be contextualised in the interview:

On the original/unfitting product:

What makes the 'unfitting product' respectively easy and difficult to market?

What are the key selling points that are typically used to promote the 'unfitting product'?

What would you have liked to be different in the 'unfitting product', to be able to sell it easier?

Why do you think people buy the 'unfitting product'?

Why do you think some people choose another product?

What do you find to be most difficult with respect to selling the product?

Is the unsuccessful good at differentiating itself from the other products on the market?

How is the 'unfitting product' positioned with respect to other products on the market?

What are the key benchmarks that make a sale in this market?

What are the expected product features in this market?

Why do you think the unsuccessful product is not successful on the market?

On the successful products on the market:

What makes the products on the market successful?

What are the key selling points in these products?

Why do you think people buy these products?

What are the successful products' differentiating factors?

What are these products' position on the market?

Why do you think people buy the successful products?

How do the successful products support the customers' personal identity?

How do the successful products support the customers functionally, emotionally and socially?

ACTION GUIDE 4:
Expert strategy 3

If your challenge involves creating a new product or product line that will strengthen the product-user fit, it may prove useful to apply expert strategy 3 (as described in Chapter 8). This action guide provides some guidelines and supporting questions to help you identify current mismatches. It also suggests how to strengthen the current product-user fit with a new and lasting solution.

Step 1: How do you identify emerging changes in users' needs, behaviours and expectations?

We recommend that your first step involves paying attention to the behaviour of users and emerging changes in this regard. What is changing? Talk to people from different environments and pay particular attention to areas in which the behaviour will manifest itself first (e.g. on social media, in types of products or services, during public debates). This focus on changes can also be supported by research regarding changes in technology, lifestyle and society.

Another way to go about this involves looking at different products (preferably outside your product category) that are gaining particular success with the user group you are targeting. These products may come from many different categories and various industries. Try to determine what kinds of changes in users' needs, behaviours or expectations the products have been driving or been particularly good at capturing.

Based on the insights you derive into the emerging changes in users' needs and behaviour, we suggest that you identify any mismatches between what the users will find attractive in the future and the products in the company's current portfolio. The aim here is to explain why the existing products will not resonate with users in the future as well as why they will not meet users' future needs and expectations. This information can be documented using worksheet 7.

Step 2: How do you reframe the strategic fit with users?

After you have identified the potential future mismatches, you should use them to reframe the direction for the new product concept. That is, you should use the mismatches as inspiration for reframing the new product's strategic fit with users.

Imagine that the new product is going to have a new identity ('It should be more like a…') and then try to conceptualise and visualise that identity.

Reframing a product cannot be done in just one attempt—it often takes many attempts to find the right reframing. As described by Schön (1983) and then nuanced by Valkenburg and Dorst (1998), you need to go through a process of naming, framing, acting and reflecting:

- First, you NAME the potential future misfit
- Then you REFRAME the new product ('The product should be like…')
- Next, you ACT—you create a concept/visualisation of the new framing
- Then you REFLECT on what you have created

When you have drawn up the first concept or visualisation of the product's reframing, you should reflect on whether the wording of the reframing is correct: Can it be more precise? Is it missing anything?

In addition, you should reflect on the visualisation of the reframing: Is the concept working? What is missing? Is there something new you have to understand about the emerging need, behaviour or expectation in question?

Based on this, you return to NAME again and you continue the loop until the reframing is in place. We have created worksheet 8 to support this reframing process.

Step 3: How do you renew the strategic fit with the market and the company?

During the process of creating the new product, it is also highly important that you consider how the new product concept can renew the strategic fit with the market and the company. As the products in the current portfolio have proved to have long-lasting product-market and product-company fits, it might be valuable to explore the company's core products. These core products are the company's most significant or iconic products. They should be studied to identify how they create a strategic fit with both the market and the company. In particular, you should look for the design principles behind these core products that create either a strong product-market fit or a strong product-company fit.

Initial core principle analysis
One way to initiate this identification of the core products' design principles is to gather all possible material regarding the products, including marketing materials, articles (newspaper/online) and product reviews, and then begin an initial identification of the design

What are the design principles behind the core products that leverage the company's core competencies and long-term strategic strengths?

What are the design principles that render the core products attractive to the company given its long-term values, purpose and culture?

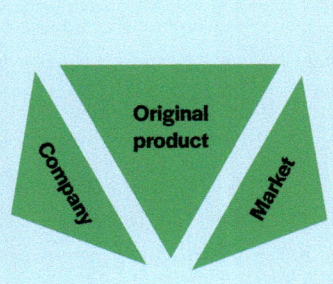

What are the design principles behind the core products that create long-term competitive advantages for the company and are unique and difficult for competitors to copy or imitate?

What are the design principles behind the core products that advance the long-term credibility of the company and, therefore, its competitive position in the market?

Figure 13.3: Model for analysing the design principles behind the company's core products

principles across these products. Figure 13.3 presents some questions that can support this analysis.

During the analysis of the core products' design principles, you should be careful not to merely identify specific features of the core product that seem to resonate with the user, the market or the company. Instead, you should try to identify the principles behind these features. Core design principles are generally supported by a number of different features and aspects.

Interviews with core stakeholders

When seeking to identify the core products' design principles, it is also a good idea to interview the designers or developers of some of the products or the marketers, sales personnel, etc., who have played a significant role in their introduction to the market. In particular, it is useful to interview people who have been with the company for a long time. Based on the interviews with these key stakeholders, it should be possible to identify the design principles behind the core products, which can otherwise be very difficult to discover. Moreover, the interview findings may add nuance to some of the core design principles you have already discovered. Worksheet 9 includes an overview of the types of questions that might prove relevant to ask stakeholders, while the design principles can be documented using worksheet 10.

Renewing core principles

The transfer and renewal of the core products' design principles to the new product must form part of the ongoing process of developing ideas, concepts and prototypes.

In accordance with the reflective practice described by Schön (1983), we suggest using the knowledge regarding the core design principles as knowledge in action. This means that the knowledge concerning the core design principles should sit in the back of your mind when you are developing ideas, concepts and prototypes. You should focus your attention on the concept or prototype you are creating and then apply the knowledge concerning the core design principles when doing so makes sense. This is known as reflection in action.

After you have finished a model or prototype, it is time to evaluate the outcome, particularly in terms of whether the core design principles have been integrated into the new concept in a new way. For instance, if you have created several concepts and you want to determine which one to move forward with, you should put together a list of core design principles and then reflect on the extent to which the different concepts implement and renew the core design principles. During this reflection, it may prove possible to come up with ways to renew some of the core principles or to combine different concepts in order to ensure an even stronger implementation.

As with the analysis, you should be very careful during the conceptual process not to simply copy features from the core products. We suggest that you focus on ensuring that all decisions and features concerning the new product become increasingly aligned with the core design principles.

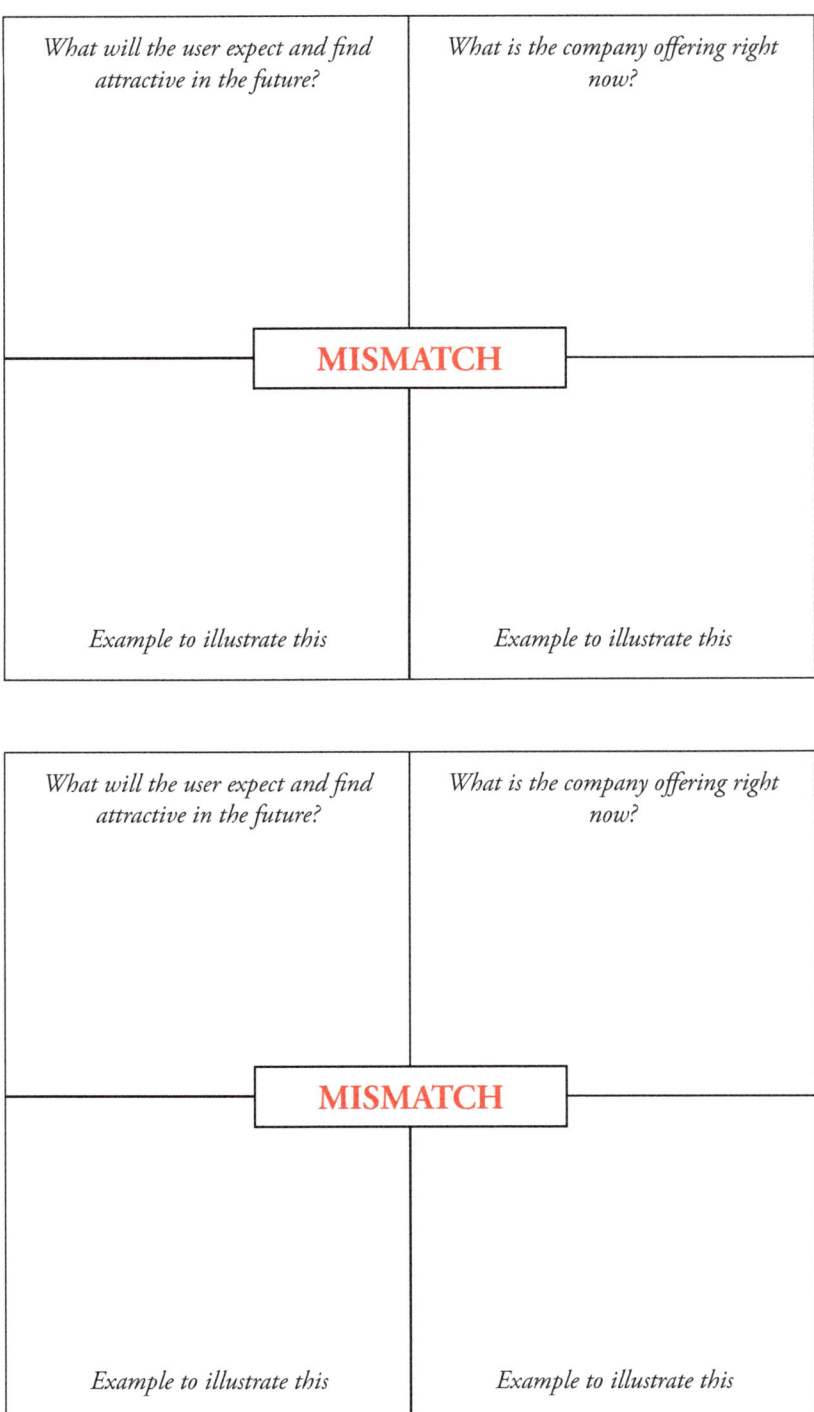

What will the user expect and find attractive in the future?	What is the company offering right now?
MISMATCH	
Example to illustrate this	*Example to illustrate this*

What will the user expect and find attractive in the future?	What is the company offering right now?
MISMATCH	
Example to illustrate this	*Example to illustrate this*

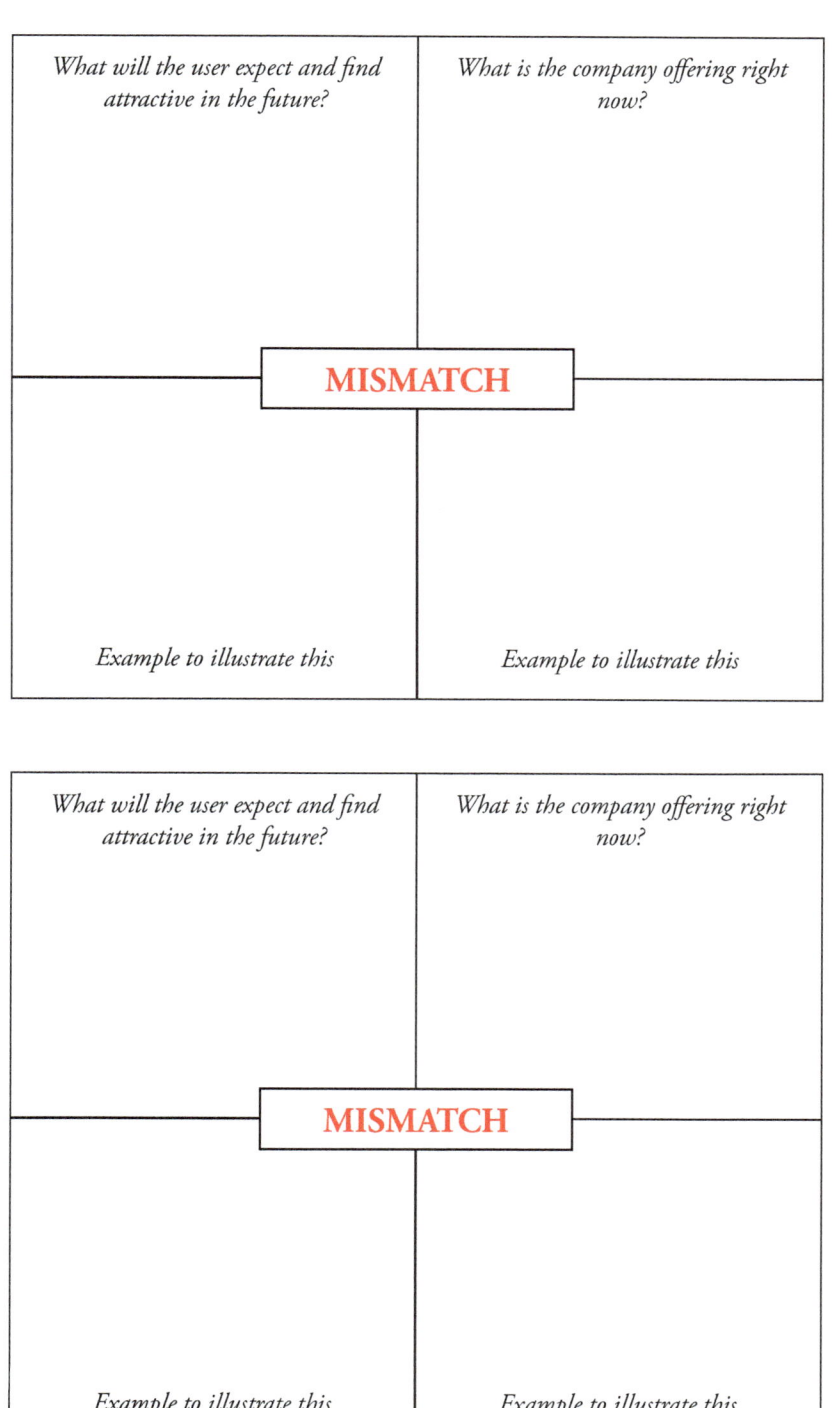

What will the user expect and find attractive in the future?	*What is the company offering right now?*
MISMATCH	
Example to illustrate this	*Example to illustrate this*

What will the user expect and find attractive in the future?	*What is the company offering right now?*
MISMATCH	
Example to illustrate this	*Example to illustrate this*

Design principle 1:

REFRAMING:

Away from:	Be like:

Design principle 2:

REFRAMING:

Away from:	Be like:

Design principle 3:

REFRAMING:

Away from:	Be like:

Design principle 4:

REFRAMING:

Away from:	Be like:

Aim of the interview

The aim of the interview with the core stakeholders is to get a deep insight into the company's products and the reasoning behind them. In order to achieve this, it is a good idea to start the interview by first going through the design process of selected products individually. This makes it possible to ask more direct and specific questions about the products, their principles and the reasoning behind them during in the interview.

Moreover, it is unlikely that the core stakeholders can articulate the design principles of the core products directly. The design principles are more likely central elements of the narratives, they tell. Once the interview is completed, it can therefore be helpful to listen to it a number of times, in order to identify the design principles. Search for the key metaphor or one-liner (e.g. 'magical interactions') and then find their connection to the strategic fit and the specific features that materialise the principle.

General questions that need to be contextualised in the interview:

The process:

What was the background/driver or initial idea behind the product?

What was the initial vision with the product?

How did the product develop over time?

What was the key insight (market, users, technology etc.) that the product is based upon?

What were the most important activities during the design process? And which important insights/findings did you get from these?

The product:

Who was the main user of the product?

Which problems, needs, challenges were the product intended to satisfy?

Why is the product experience/expression and interaction the way it is?

Is there any new technology or new use of technology in the product?

How did the product differentiate from other products on the market when it was first introduced?

How was the product positioned compared to its competitors?

Which of the company's core competencies and strategic strengths did the product built upon?

Why did the company management find the product attractive?

The reasoning:

What are the key principles in the product? Where did these come from?

Were there any disagreements with respect to specific parts of the product?

What kind of things could the product never do? And why?

Were there any features or aspects that were deselected in the product? And why?

What kinds of changes were made to the product from concept to production? And why?

Were there any conflicting insights or information you had to deal with?

Product-company fit

Design principle 1:

Strategic fit with the company

Solution elements of the core products

Design principle 2:

Strategic fit with the company

Solution elements of the core products

Design principle 3:

Strategic fit with the company

Solution elements of the core products

Product-market fit

Design principle 4:

Strategic fit with the market

Solution elements of the core products

Design principle 5:

Strategic fit with the market

Solution elements of the core products

Design principle 6:

Strategic fit with the market

Solution elements of the core products

ACTION GUIDE 5:
Expert strategy 4

If your challenge involves creating a new product that will strengthen the product-market fit, it will likely prove useful to apply strategy 4 (as described in Chapter 9). This action guide provides some guidelines and supporting questions to help you to identify potential extensions of the product value. It also suggests how to strengthen the current product-market fit with a new and lasting solution.

Step 1: How do you identify ways to extend the product value?

We suggest that your first step involves investigating the user's relation to the company's products. The aim behind doing so is to understand what motivates the user to use, keep, maintain and repair the product. Likewise, the user research should also disclose the points in the use journey where the user experiences hesitation, frustration or concern.

Through mapping the existing user experience, it is possible to identify functional, personal and social values that the product might be missing. It might also prove relevant to investigate different stakeholders' 'experience' (e.g. installers, retailers, service providers, sales personnel or other people who interact with the product, besides the user or purchasing customer) and determine if there is any potential in extending the value for them.

There are various different types of user research methodologies. It is not important which one you choose, although it is important that you consider both functional and emotional value in the mapping and that all potential stakeholders are considered. Moreover, we recommend that you create a user experience map and use both storytelling and visualisation to illustrate the user's relationship with the product, the brand and the company over time.

Based on the user research mapping, it should be possible to identify potential areas where new functional, personal or social value could be added to the product's user experience. These potentials can be recorded using worksheet 11, where it is also possible to show how the values provide the product with long-term competitive advantages in the market.

Step 2: How do you reframe the strategic fit with the market?

Once you have identified ways to extend the product value, you can use the knowledge concerning the potential to reframe the new product's long-lasting strategic fit with the market.

Imagine that the new product is having a new identity ('It should be more like a...') and then try to conceptualise and visualise that identity.

Reframing a product cannot be done in just one attempt – it often takes many attempts to find the correct reframing. As described by Schön (1983) and then nuanced by Valkenburg and Dorst (1998), you need to go through a process of naming, framing, acting and reflecting:

- First, you NAME the potential for extending product value
- Then you REFRAME the new product ('the product should be like…')
- Next, you ACT—you create a concept/visualisation of the new product with the new framing
- Then you REFLECT on what you have created

After you have drawn up the first concept or visualisation of the product's reframing, you should reflect on whether the wording of the reframing is correct: Can it be more precise? Is it missing anything?

In addition, you should reflect on the visualisation of the reframing: Is the concept working? What is missing? Is there something new you have to understand about the extension of the value in order to proceed?

Based on the reflection, you should return to NAME again and continue the loop until the reframing is in place. We have created worksheet 12 to support you with the reframing process.

Step 3: How do you realign the new reframing with the current strategic fit?

During the process of creating the new product, it is also highly important to consider how the new product can create a long-lasting strategic fit with the user and the company. As the products in the company's current portfolio have created long-term product-user and product-company fits, it might be valuable to explore the company's core products. These core products are the company's most significant or iconic products. They should be studied to identify how they create a strategic fit with both the user and the company. In particular, you should look for the design principles behind these 'core products' that create either a lasting product-user fit or a lasting product-company fit.

What are the design principles behind the core products that solve users' long-lasting problems and address their long-term needs, wishes and aspirations?

What are the design principles behind the core products that leverage the company's core competencies and long-term strategic strengths?

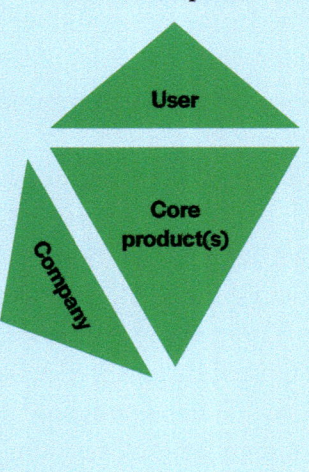

What are the design principles that render the core products attractive to the company given its long-term values, purpose and culture?

Figure 13.4: Model for analysing the design principles behind the company's core products

Initial core principle analysis

One way to initiate the identification of the core principles is to gather all possible material concerning the selected products, including marketing materials, articles (newspaper/online) product reviews, and then begin an initial identification of the design principles across these products. Figure 13.4 presents some questions that can support this analysis, while worksheet 13 can be used to document the core design principles.

During the analysis of the core products' design principles, be careful not to simply identify specific features that appear to resonate with the user or company. Rather, look for the principles behind such features. Design principles are generally supported by several different features and aspects.

Interviews with core stakeholders

During the quest to identify design principles, it is also a good idea to interview the designers or developers of some of the core products or the marketers, sales personnel, strategic managers, etc. who have been heavily involved with them. It is particularly useful to interview people who have been with the company for a long time. Based on the interviews with these key stakeholders, it should be possible to identify the design principles behind the core products, which could otherwise be very difficult to discover. Moreover, the interview data may add nuance to some of the core principles you have already discovered. Worksheet 14 provides an overview of the types of questions that could be relevant to ask stakeholders.

Renewing core principles

The transfer and renewal of the core products' design principles to the new product must form part of the ongoing process of developing ideas, concepts and prototypes.

In accordance with the reflective practice described by Schön (1983), we suggest using the knowledge regarding the core design principles as knowledge in action. This means that the knowledge concerning the core design principles should sit in the back of your mind when you are developing ideas, concepts and prototypes. You should focus your attention on the concept or prototype you are creating and then apply the knowledge concerning the core design principles when doing so makes sense. This is known as reflection in action.

Once you have finished a model or a prototype, you should evaluate whether the core principles have been transferred and renewed. For instance, if you have created several concepts and you want to determine which one to move forward with, you should put together a list of core design principles and then reflect on the extent to which the different concepts implement and renew the core principles. During this reflection, it may be possible to come up with ways to renew some of the core principles or combine different concepts to render the implementation stronger.

As with the analysis, you should be very careful during the conceptual process not to just copy features from the core products. Instead, focus on ensuring that all decision and features concerning the new product become increasingly aligned with the core design principles.

What characterises the customer experience now?	*In which way can the customer experience be improved?*
	DIFFERENTIATION
Example to illustrate this	*Examples of products or services that create this experience*

What characterises the customer experience now?	*In which way can the customer experience be improved?*
	DIFFERENTIATION
Example to illustrate this	*Examples of products or services that create this experience*

What characterises the customer experience now?	In which way can the customer experience be improved?
DIFFERENTIATION	
Example to illustrate this	Examples of products or services that create this experience

What characterises the customer experience now?	In which way can the customer experience be improved?
DIFFERENTIATION	
Example to illustrate this	Examples of products or services that create this experience

---

Design principle 1:

REFRAMING:

Away from:	Be like:

Design principle 2:

REFRAMING:

Away from:	Be like:

Design principle 3:

REFRAMING:

Away from:	Be like:

Design principle 4:

REFRAMING:

Away from:	Be like:

Product-company fit

Design principle 1:

Strategic fit with the company

Solution elements of the core products

Design principle 2:

Strategic fit with the company

Solution elements of the core products

Design principle 3:

Strategic fit with the company

Solution elements of the core products

Product-user fit

Design principle 4:

Strategic fit with the user

Solution elements of the core products

Design principle 5:

Strategic fit with the user

Solution elements of the core products

Design principle 6:

Strategic fit with the user

Solution elements of the core products

Aim of the interview

The aim of the interview with the core stakeholders is to get a deep insight into the company's products and the reasoning behind the products. In order to achieve this, it is a good idea to start the interview by first going through the design process of selected products individually. This makes it possible to ask more direct and specific questions about the products, their principles and the reasoning behind them during in the interview.

Moreover, it is unlikely that the core stakeholders can articulate the design principles of the core products directly. The design principles are more likely central elements of the narratives, they tell. Once the interview is completed, it can therefore be helpful to listen to it a number of times, in order to identify the design principles. Search for the key metaphor or one-liner (e.g 'security first') and then find their connection to the strategic fit and the specific features that materialise the principle.

General questions that need to be contextualised in the interview:

The process:

What was the background/driver or initial idea behind the product?

What was the initial vision with the product?

How did the product develop over time?

What was the key insight (market, users, technology etc.) that the product is based upon?

What were the most important activities during the design process? And which important insights/findings did you get from these?

The product:

Who was the main user of the product?

Which problems, needs, challenges were the product intended to satisfy?

Why is the product experience/expression and interaction the way it is?

Is there any new technology or new use of technology in the product?

How did the product differentiate from other products on the market, when it was first introduced?

How was the product positioned compared to its competitors?

Which of the company's core competencies and strategic strengths did the product built upon?

Why did the company management find the product attractive?

The reasoning:

What are the key principles in the product? Where did they come from?

Were there any disagreements with respect to specific parts of the product?

What kind of things could the product never do? And why?

Were there any features or aspects that were deselected in the product? And why?

What kinds of changes were made to the product from concept to production? And why?

Were there any conflicting insights or information you had to deal with?

ACTION GUIDE 6:
Expert strategy 5

If your challenge involves creating a strategically durable product for an entirely new market and user segment, it may prove useful to apply expert strategy 5 (as described in Chapter 10). This action guide will help you to frame lasting strategic fits with the user, the market and the company from scratch.

Step 1: How do you frame a product-user fit from scratch?

We suggest that you begin the process with a fast loop of collecting all available material concerning the new user group. Even if the users might appear somewhat familiar to you, you should aspire to see and understand them from a fresh perspective. You should be curious about what kinds of lives they live, what is important to them and what kinds of products really resonate with them.

As soon as possible (preferably within the first couple of weeks), work with this new user group and develop 5–10 very different and explorative ideas/concepts. Present these ideas/concepts to two or three users who are representative of the user group. The aim is not to present ideas or concepts that can be adjusted in further designs; rather, the idea is to explore different design directions and identify which of them resonate with the users and hook their attention. Moreover, this exploration might reveal the needs or aspirations that are key to creating a successful and long-lasting product-user fit.

The next stage involves identifying inspirational products that have proven to be particularly successful at addressing some of the long-lasting needs or aspirations of the user group. During this process, you should search for the key design principles that cause the inspirational products to resonate with users and create a long-term strategic fit. The design principles behind the inspirational products can be documented using worksheet 15.

Once the key design principles have been identified, a new loop of developing product ideas and concepts should be initiated. This time, the idea and concept generation aims to incorporate some of the design principles behind the inspirational products. The ideas and concepts should be continuously tested with the new users in order to identify which ones 'hook' the users' attention and which do not. We suggest that you perform several loops of creating, testing with users, recreating and testing with users until a concept proves to create a lasting product-user fit.

Step 2: How do you frame a product-market fit from scratch?

The second step in this strategy ensures that the product concept you are creating will also hold a lasting competitive position in the market. Similar to the engagement with users, it is important to see the new market through fresh eyes and be careful not to adopt any previous beliefs or assumptions based on the company's existing markets.

Instead, explore the new market, look at key competitors and identify their long-term competitive advantages. In particular, it is useful to identify the most successful and long-lasting products on the new market. You should identify the design principles that have been the drivers of this long-term competitiveness. The design principles behind the inspirational products can be documented using worksheet 16.

Next, we advise you to develop a new set of ideas and concepts based on the knowledge concerning the key design principles that create a long-lasting strategic fit with the new market. It is a good idea to continuously test the developed ideas with customers, sales personnel, marketers, etc. to determine what hooks their attention. We suggest that you perform several loops of creating, testing on the market, recreating and testing on the market until the product concept proves to create a lasting product-market fit.

Step 3: How do you frame the product-company fit from scratch?

The third step in developing a new product for an entirely new market involves ensuring that the new product will have a lasting strategic fit with the company. To achieve this, it is vital to continuously present the concepts and ideas to internal stakeholders in order to determine their likes, dislikes, concerns, hooks and objections. This will also help you to become increasingly aware of which design principles behind the product concept create a lasting strategic fit with the company. In many ways, the product-company fit needs to be negotiated and all objections need to be heard. Entering a new market means that it is often not possible to renew or adopt a strategic fit from one of the company's core products. Instead, the strategic fit with the company needs to be framed from scratch, and it is up to you and the design team to ensure alignment.

Design principle 1:

Inspirational products	Solution principles

Design principle 2:

Inspirational products	Solution principles

Design principle 3:

Inspirational products	Solution principles

Design principle 1:

Inspirational products	Solution principles

Design principle 2:

Inspirational products	Solution principles

Design principle 3:

Inspirational products	Solution principles

Bibliography

Adamson, Glenn. 2003. *Industrial Strength Design: How Brooks Stevens Shaped Your World.* Cambridge: The MIT Press.

Alqahtani, Ammar Y. and Surendra M. Gupta. 2017. "Warranty as a Marketing Strategy for Remanufactured Products." *Journal of Cleaner Production* 161: 1294–1307. doi:https://doi.org/10.1016/j.jclepro.2017.06.193. https://www.sciencedirect.com/science/article/pii/S095965261731363X.

Andreasen, Mogens Myrup, Claus Thorp Hansen, and Philip Cash. 2015. *Conceptual Design: Interpretations, Mindset and Models.* Switzerland: Springer.

Antikainen, Maria and Katri Valkokari. 2016. "A Framework for Sustainable Circular Business Model Innovation." *Technology Innovation Management Review* 6 (7) (Jul 27): 5–12. doi:10.22215/timreview/1000. https://search.proquest.com/docview/1963328830.

Aziz, N. A., D. A. Wahab, R. Ramli, and C. H. Azhari. 2016. "Modelling and Optimisation of Upgradability in the Design of Multiple Life Cycle Products: A Critical Review." *Journal of Cleaner Production* 112: 282–290. doi:https://doi.org/10.1016/j.jclepro.2015.08.076. https://www.sciencedirect.com/science/article/pii/S0959652615011671.

Bakker, Conny, Feng Wang, Jaco Huisman, and Marcel den Hollander. 2014. "Products that Go Round: Exploring Product Life Extension through Design." *Journal of Cleaner Production* 69 (Apr 15): 10–16. doi:10.1016/j.jclepro.2014.01.028. https://dx.doi.org/10.1016/j.jclepro.2014.01.028.

Bakker, C., M. den Hollander, E. van Hinte, and Y. Zljlstra. 2019. *Products that Last – Product Design for Circular Business Models.* Amsterdam: BIS Publishers.

Bakker, C. A., R. Mugge, C. Boks, and M. Oguchi. 2021. "Understanding and Managing Product Lifetimes in Support of a Circular Economy." *Journal of Cleaner Production* 279 (Jan 10): 123764. doi:10.1016/j.

jclepro.2020.123764. https://dx.doi.org/10.1016/j.jclepro.2020.123764.

Bayus, Barry L. 1991. "The Consumer Durable Replacement Buyer." *Journal of Marketing* 55 (1): 42–51. doi:10.1177/002224299105500104. https://www.jstor.org/stable/1252202.

Besch, Katrin. 2005. "Product-Service Systems for Office Furniture: Barriers and Opportunities on the European Market." *Journal of Cleaner Production* 13 (10–11): 1083–1094.

Bijen, J. 2006. *Durability of Engineering Structures: Design, Repair and Maintenance.* Cambridge, UK: Woodhead.

Bocken, N. M. P., S. W. Short, P. Rana, and S. Evans. 2014. "A Literature and Practice Review to Develop Sustainable Business Model Archetypes." *Journal of Cleaner Production* 65 (Feb 15): 42–56. doi:10.1016/j. jclepro.2013.11.039. https://dx.doi.org/10.1016/j.jclepro.2013.11.039.

Bocken, N. M. P., P. Rana, and S. W. Short. 2015. "Value Mapping for Sustainable Business Thinking." *Journal of Industrial and Production Engineering* 32 (1) (Jan 02): 67–81. doi:10.1080/21681015.2014.10003 99. http://www.tandfonline.com/doi/abs/10.1080/21681015.2014.10003 99.

Bocken, N. M. P. and S. W. Short. 2016. "Towards a Sufficiency-Driven Business Model: Experiences and Opportunities." *Environmental Innovation and Societal Transitions* 18 (Mar): 41–61. doi:10.1016/j.eist.2015.07.010. https://dx.doi.org/10.1016/j.eist.2015.07.010.

Bocken, Nancy M. P., Ingrid De Pauw, Conny Bakker, and Bram Van Der Grinten. 2016. *Product Design and Business Model Strategies for a Circular Economy.* Vol. 33 Informa UK Limited. doi:10.1080/21681015.2016.117 2124.

BrickEconomy. 2022. "A Comprehensive Guide to LEGO Economics, Market Values and Trends." BrickEconomy.com.2022, BrickEconomy. com.

Brown, Tim. 2009. *Change by Design: How Design Thinking Transforms Organizations and Inspires Innovation.* New York: HarperCollins Publishers.

Buchanan, Richard. 1992. "Wicked Problems in Design Thinking." *Design Issues* 8 (2): 5–21.

Buijs, Jan. 2012. *The Delft Innovation Method: A Design Thinker's Guide to Innovation.* The Hague: Eleven International Publishing.

Burns, B. 2010. "Re-Evaluating Obsolescence and Planning for It." In *Longer Lasting Products: Alternatives to the Throwaway Society*, edited by Tim Cooper. Farnham: Gower Publishing.

Calabretta, Giulia, Gerda Gemser, and Ingo Karpen. 2016. *Strategic Design: 8 Essential Practices Every Strategic Designer must Master.* Amsterdam: BIS Publishers.

Chapman, Jonathan. 2005. *Emotionally Durable Design: Objects, Experiences and Empathy.* London: Earthscan.

Chapman, Jonathan. 2009. "Design for (Emotional) Durability." *Design Issues* 25 (4): 29–35.

Chapman, Jonathan. 2021. *Meaningful Stuff – Design that Lasts.* London: MIT Press.

Collins, James C. and Jerry I. Porras. 1996. "Building Your Company's Vision." *Harvard Business Review* 74 (5).

Cooper, Tim. 1994. "The Durability of Consumer Durables." *Business Strategy and the Environment* 3 (1): 23–30.

Cooper, Tim. 2004. "Inadequate Life? Evidence of Consumer Attitudes to Product Obsolescence." *Journal of Consumer Policy* 27: 421–449.

Cooper, Tim. 2010. "The Significance of Product Longevity." In *Longer Lasting Products: Alternatives to the Throwaway Society*, edited by Tim Cooper, 3-36. Farnham: Gower Publishing.

Curedale, R. 2013. *Design Thinking: Process and Methods Manual.* Design Community College.

de Mozota, Brigitte Borja. 1998. "Structuring Strategic Design Management: Michael Porter's Value Chain." *Design Management Journal* (Former Series) 9 (2): 26–31.

Den Hollander, Marcel C., Conny A. Bakker, and Erik Jan Hultink. 2017. "Product Design in a Circular Economy: Development of a Typology of Key Concepts and Terms." *Journal of Industrial Ecology* 21 (3): 517–525.

Dorst, Kees. 2015. *Frame Innovation: Create New Thinking by Design.* Massachusetts Institute of Technology.

Dorst, Kees and Nigel Cross. 2001. "Creativity in the Design Process: Co-Evolution of Problem–Solution." *Design Studies* 22 (5): 425–437.

Ellen MacArthur Foundation. 2013. *Towards the Circular Economy – Economic and Business Rationale for an Accelerated Transition.* Ellen MacArthur Foundation Publishing.

Ellen MacArthur Foundation. "Completing the Picture – how the Circular Economy Tackles Climate Change." 2021, https://ellenmacarthurfoundation.org/completing-the-picture.

Ertz, Myriam, Sébastien Leblanc-Proulx, Emine Sarigöllü, and Vincent Morin. 2019. "Made to Break? A Taxonomy of Business Models on Product Lifetime Extension." *Journal of Cleaner Production* 234: 867–880.

European Commission. 2020. *Circular Economy Action Plan For a Cleaner and More Competitive Europe.* Luxembourg: Publications Office of the European Union.

European Parliament. 2015. "Circular Economy: The Importance of Re-using Products and Materials." https://www.europarl.europa.eu/.2021, https://www.europarl.europa.eu/news/en/headlines/economy/20150701STO72956/circular-economy-the-importance-of-re-using-products-and-materials.

Faber, N. R., R. J. Jorna, and J. M. L. Van Engelen. 2009. "The Sustainability of "sustainability" – a Study into the Conceptual Foundations of the Notion of "sustainability"." *Journal of Environmental Assessment Policy and Management* 7 (1): 1–33. doi:10.1142/S1464333205001955. https://

www.jstor.org/stable/enviassepolimana.7.1.1.

Flyvbjerg, Bent. 2006. "Five Misunderstandings about Case-Study Research." *Qualitative Inquiry* 12 (2): 219–245.

Geng, Yong and Brent Doberstein. 2008. "Developing the Circular Economy in China: Challenges and Opportunities for Achieving 'Leapfrog Development'." *The International Journal of Sustainable Development & World Ecology* 15 (3): 231–239.

Granberg, Bjorn. 1997. "The Quality Re-Evaluation Process: Product Obsolescence in a Consumer-Producer Interaction Framework." University of Stockholm.

Gupta, Diwakar and Yigal Gerchak. 1995. "Joint Product Durability and Lot Sizing Models." *European Journal of Operational Research* 84 (2): 371–384. doi:https://doi.org/10.1016/0377-2217(93)E0273-Z. https://www.sciencedirect.com/science/article/pii/0377221793E0273Z.

Haapala, Karl R., Kari L. Brown, and John W. Sutherland. 2008. "A Life Cycle Environmental and Economic Comparison of Clothes Washing Product-Service Systems." *Transactions of NAMRI/SME* 36: 333-340.

Haase, Louise Møller and Linda Nhu Laursen. 2019. "Meaning Frames: The Structure of Problem Frames and Solution Frames." *Design Issues* 35 (3) (Jul 01): 20–34. doi:10.1162/desi_a_00547. http://www.mitpressjournals.org/doi/abs/10.1162/desi_a_00547.

Haffmans, S., M. van Gelder, E. van Hinte, and Y. Zljlstra. 2018. *Products that Flow – Circular Business Models and Design Strategies for Fast-Moving Consumer Goods.* Amsterdam: BIS Publishers.

Haines-Gadd, Merryn, Jonathan Chapman, Peter Lloyd, Jon Mason, and Dzmitry Aliakseyeu. 2018. "Emotional Durability Design Nine—A Tool for Product Longevity." *Sustainability (Basel, Switzerland)* 10 (6) (Jun 11): 1948. doi:10.3390/su10061948. https://explore.openaire.eu/search/publication?articleId=dedup_wf_001::8b71e33bf14d79b658bdd09434377ace.

Hankammer, Stephan, Sebastian Brenk, Hannah Fabry, Anne Nordemann, and Frank T. Piller. 2019. "Towards Circular Business Models: Identifying Consumer Needs Based on the Jobs-to-be-done Theory." *Journal of Cleaner Production* 231: 341–358.

Hedstrom, Gilbert. 2019. *Sustainability – What It Is and How to Measure It.* Berlin: De Gruyter.

Hey, Jonathan H. G., Caneel K. Joyce, and Sara L. Beckman. 2007. "Framing Innovation: Negotiating Shared Frames during Early Design Phases." *Journal of Design Research* (Geneva, Switzerland) 6 (1–2) (Jan 01): 79-99. doi:10.1504/JDR.2007.015564. https://www.inderscienceonline.com/doi/10.1504/JDR.2007.015564.

Holston, David. 2011. *The Strategic Designer: Tools & Techniques for Managing the Design Process.* Cincinnati: Simon and Schuster.

Hooley, Graham, Brigitte Nicoulaud, John Rudd, and Nick Lee. 2020. *Marketing Strategy and Competitive Positioning.* 7th ed. Harlow: Pearson.

IPCC. 2014. "Summary for Policymakers." In *Climate Change 2014: Mitigation of Climate Change. Contribution of Working Group III to the Fifth Assessment Report of the Intergovernmental Panel on Climate Change*, edited by Edenhofer, O., R. Pichs-Madruga, Y. Sokona, E. Farahani, S. Kadner, K. Seyboth, A. Adler, I. Baum, S. Brunner, P. Eickemeier, B. Kriemann, J. Savolainen, S. Schlömer, C. von Stechow, T. Zwickel, and J.C. Minx. Cambridge, UK and New York: Cambridge University Press.

Jensen, Peter Byrial, Linda Nhu Laursen, and Louise Møller Haase. 2021. "Barriers to Product Longevity: A Review of Business, Product Development and User Perspectives." *Journal of Cleaner Production* 313 (Sep 01): 127951. doi:10.1016/j.jclepro.2021.127951. https://dx.doi.org/10.1016/j.jclepro.2021.127951.

Johnson, Gerry, Kevan Scholes, and Richard Whittington. 2008. *Exploring Corporate Strategy: Text and Cases.* Harlow: Pearson Education.

Keoleian, G. A. and D. Menery. 1993. *Life Cycle Design Guidance Manual.* EPA Washington, DC, US Government Publishing Office.

Khan, Muztoba Ahmad, Sameer Mittal, Shaun West, and Thorsten Wuest. 2018. "Review on Upgradability – A Product Lifetime Extension Strategy in the Context of Product Service Systems." *Journal of Cleaner Production* 204: 1154–1168. doi:https://doi.org/10.1016/j.jclepro.2018.08.329. https://www.sciencedirect.com/science/article/pii/S095965261832691X.

Kim, Hyung Chul, Gregory A. Keoleian, and Yuhta A. Horie. 2006. "Optimal Household Refrigerator Replacement Policy for Life Cycle Energy, Greenhouse Gas Emissions, and Cost." *Energy Policy* 34 (15): 2310–2323.

Korhonen, J., Honkasalo, A., & Seppälä, J. 2018. "Circular Economy: The concept and its limitations." *Ecological Economics* 143 (2018), 37–46. https://10.1016/j.ecolecon.2017.06.041

Kotler, Philip. 2003. *Marketing Management.* New York: Prentice Hall.

Krippendorff, Klaus. 2006. *The Semantic Turn: A New Foundation for Design.* Boca Raton: Routledge.

Kumar, Rajesh. 2014. "Managing Ambiguity in Strategic Alliances." *California Management Review* 56 (4): 82–102.

Kuo, Tsai Chi. 2011. "Simulation of Purchase or Rental Decision-Making Based on Product Service System." *The International Journal of Advanced Manufacturing Technology* 52 (9): 1239–1249.

Laurel, Brenda. 2003. *Design Research: Methods and Perspectives.* MIT Press.

Laursen, Linda Nhu and Louise Møller Haase. 2019. "The Shortcomings of Design Thinking when Compared to Designerly Thinking." *The Design Journal* 22 (6) (Nov 02): 813–832. doi:10.1080/14606925.2019.1652531. http://www.tandfonline.com/doi/abs/10.1080/14606925.2019.1652531.

LEGO. 2022. "Environment – Reporting – Sustainability." https://www.lego.com/sustainability/reporting/environment/.

Lewandowski, Mateusz. 2016. "Designing the Business Models for Circular economy—Towards the Conceptual Framework." *Sustainability* 8 (1): 43.

Lockwood, T. and T. Walton. 2008. *Building Design Strategy: Using Design to Achieve Key Business Objectives.* New York: Allworth.

Lofthouse, V. A. and S. Prendeville. 2017. "Considering the User in the Circular Economy." Delft University of Technology and IOS Press, 8–10 November 2017.

Magretta, J. 2002. "Why Business Models Matter." *Harvard Business Review* 80: 86–92.

Manninen, Kaisa, Sirkka Koskela, Riina Antikainen, Nancy Bocken, Helena Dahlbo, and Anna Aminoff. 2017. "Do Circular Economy Business Models Capture Intended Environmental Value Propositions?" *Journal of Cleaner Production* 171: 413–422. doi:10.1016/j.jclepro.2017.10.003. https://dx.doi.org/10.1016/j.jclepro.2017.10.003.

Martin, Roger. 2009. *The Design of Business: Why Design Thinking is the Next Competitive Advantage.* Boston: Harvard Business Review Press.

McDonough, William and Michael Braungart. 2002. *Cradle to Cradle – Remaking the Way We Make Things.* New York: North Point Press.

Merholz, Peter, Todd Wilkens, Brandon Schauer, and David Verba. 2008. *Subject to Change: Creating Great Products and Services for an Uncertain World-Adaptive Path on Design.* O'Reilly Media.

Møller, Louise and Christian Tollestrup. 2013. *Creating Shared Understanding in Product Development Teams: How to 'Build the Beginning.'* London: Springer.

Morelli, Nicola. 2006. "Developing new product service systems (PSS): Methodologies and operational tools." *Journal of Cleaner Production,* 14(17) 1495–1501.

Moreno, Mariale, Carolina De los Rios, Zoe Rowe, and Fiona Charnley. 2016. "A Conceptual Framework for Circular Design." *Sustainability (Basel, Switzerland)* 8 (9) (Sep 13): 937. doi:10.3390/su8090937. https://explore.openaire.eu/search/publication?articleId=dedup_wf_001::21ddfdb6fe486fd1a618667579f728d6.

Mugge, Ruth, Jan P. L. Schoormans, and Hendrik N. J. Schifferstein. 2005. "Design Strategies to Postpone Consumers' Product Replacement: The Value of a Strong Person-Product Relationship." *The Design Journal* 8 (2): 38–48.

Nixon, Natalie W. 2016. *Strategic Design Thinking: Innovation in Products, Services, Experiences, and Beyond.* London: Fairchild Books.

Norman, Donald A. 2004. *Emotional Design: Why We Love (Or Hate) Everyday Things.* New York: Basic Books.

Nußholz, Julia L. K. 2018. "A Circular Business Model Mapping Tool for Creating Value from Prolonged Product Lifetime and Closed Material Loops." *Journal of Cleaner Production* 197 (Oct 01): 185–194. doi:10.1016/j. jclepro.2018.06.112. https://dx.doi.org/10.1016/j.jclepro.2018.06.112.

Oghazi, Pejvak and Rana Mostaghel. 2018. "Circular Business Model Challenges and Lessons Learned—An Industrial Perspective." *Sustainability* 10 (3): 739.

Osterwalder, A., Pigneur, Y., Bernada, G., Smith, A. 2014. *Value Proposition Design.* New Jersey: John Wiley & Sons.

Oswald, Irina and Armin Reller. 2011. "E-Waste: A Story of Trashing, Trading, and Valuable Resources." *GAIA-Ecological Perspectives for Science and Society* 20 (1): 41–47.

Pangburn, Michael S. and Euthemia Stavrulaki. 2014. "Take Back Costs and Product Durability." *European Journal of Operational Research* 238 (1) (Oct 01): 175–184. doi:10.1016/j.ejor.2014.03.008. https://dx.doi. org/10.1016/j.ejor.2014.03.008.

Penty, Jane. 2020. *Product Design and Sustainability – Strategies, Tools and Practice.* New York: Routledge.

Poppelaars, Flora, Conny Bakker, and Jo van Engelen. 2018. "Does Access Trump Ownership? Exploring Consumer Acceptance of Access-Based Consumption in the Case of Smartphones." *Sustainability (Basel, Switzerland)* 10 (7) (Jul): 2133. doi:10.3390/su10072133. https://www.narcis.nl/ publication/RecordID/oai:pure.rug.nl:publications%2F2e9cddb7-9efc-

4407-a4d9-1cd5fdb6a135.

Porter, Michael. 1980. *Competitive Strategy: Techniques for Analyzing Industries and Competitors.* New York: Free Press.

Richardson, J. 2008. "The Business Model: An Integrative Framework for Strategy Execution." *Strategic Change* 17 (5–6): 133–144. doi:10.1002/jsc.821.

Rivera, Julio L. and Amrine Lallmahomed. 2016. "Environmental Implications of Planned Obsolescence and Product Lifetime: A Literature Review." *International Journal of Sustainable Engineering* 9 (2): 119–129.

Salvador, Rodrigo, Murillo Vetroni Barros, Leila Mendes da Luz, Cassiano Moro Piekarski, and Antonio Carlos de Francisco. 2020. "Circular Business Models: Current Aspects that Influence Implementation and Unaddressed Subjects." *Journal of Cleaner Production* 250: 119555.

Sanders, Elizabeth B. -N. and Pieter Jan Stappers. 2014. "Probes, Toolkits and Prototypes: Three Approaches to Making in Codesigning." *CoDesign* 10 (1) (Jan 02): 5–14. doi:10.1080/15710882.2014.888183. http://www.tandfonline.com/doi/abs/10.1080/15710882.2014.888183.

Schifferstein, Hendrik and Elly Zwartkruis-Pelgrim. 2008. "Consumer-Product Attachment: Measurement and Design Implications." *International Journal of Design* 2 (3).

Schön, Donald. 1983. *The Reflective Practitioner.* New York: Basic Books.

Schön, Donald and Martin Rein. 1994. *Frame Reflection: Toward the Resolution of Intractable Policy Controversies.* New York: Basic Books.

Sinclair, Matt, Leila Sheldrick, Mariale Moreno, and Emma Dewberry. 2018. "Consumer Intervention Mapping—A Tool for Designing Future Product Strategies within Circular Product Service Systems." *Sustainability* 10 (6): 2088.

Snowden, David and Mary Boone. 2007. "A Leader's Framework for Decision Making." *Harvard Business Review* (November).

Stahel, Walter R. 1982. "The Product Life Factor." In *An Inquiry into the Nature of Sustainable Societies: The Role of the Private Sector* (Series: 1982 Mitchell Prize Papers), NARC, edited by Susan Grinton, 72–96. Houston: HARC.

Stahel, Walter. 2010. *The Performance Economy.* London: Springer.

Stahel, Walter. 2016. "The Circular Economy." *Nature* (531): 435–438.

Stahel, Walter. 2019. *The Circular Economy – A User's Guide.* New York: Routledge.

Stompff, Guido, Frido Smulders, and Lilian Henze. 2016. "Surprises are the Benefits: Reframing in Multidisciplinary Design Teams." *Design Studies* 47: 187–214.

Tam, Edwin, Katie Soulliere, and Susan Sawyer-Beaulieu. 2019. "Managing Complex Products to Support the Circular Economy." *Resources, Conservation and Recycling* 145: 124–125. doi:https://doi.org/10.1016/j.resconrec.2018.12.030. https://www.sciencedirect.com/science/article/pii/S0921344918304865.

Tukker, Arnold. 2015. "Product Services for a Resource-Efficient and Circular Economy – a Review." *Journal of Cleaner Production* 97 (Jun 15): 76–91. doi:10.1016/j.jclepro.2013.11.049. https://dx.doi.org/10.1016/j.jclepro.2013.11.049.

Ulwick, Anthony. 2005. *What Customers Want: Using Outcome-Driven Innovation to Create Breakthrough Products and Services.* New York: McGraw-Hill.

Valkenburg, R., & Dorst, K. 1998. "The Reflective Practice of Design Teams." *Design Studies*, 19(3), 249–271.

Valkenburg, R. 2000. "The Reflective Practice in Product Design Teams." Delft University of Technology.

Van Doorsselaer, Karine and Rudolf Koopmans. 2021. *Ecodesign: A Life Cycle Approach for a Sustainable Future.* Munich: Carl Hanser Verlag.

Van Loon, Patricia, Derek Diener, and Steve Harris. 2020. "Circular Products and Business Models and Environmental Impact Reductions: Current Knowledge and Knowledge Gaps." *Journal of Cleaner Production* 288. doi:10.1016/j.jclepro.2020.125627. https://dx.doi.org/10.1016/j.jclepro.2020.125627.

Van Nes, Nicole. 2003. "Replacement of Durables: Influencing Product Lifetime through Product Design." Erasmus University Rotterdam.

Van Nes, Nicole and Jacqueline Cramer. 2005. "Influencing Product Lifetime through Product Design." *Business Strategy and the Environment* 14 (5) (Sep): 286–299. doi:10.1002/bse.491. https://api.istex.fr/ark:/67375/WNG-2CSJ2CH5-5/fulltext.pdf.

Van Nes, Nicole and Jacqueline Cramer. 2006. "Product Lifetime Optimization: A Challenging Strategy Towards More Sustainable Consumption Patterns." *Journal of Cleaner Production* 14 (15): 1307–1318. doi:10.1016/j.jclepro.2005.04.006. https://dx.doi.org/10.1016/j.jclepro.2005.04.006.

Verganti, Roberto. 2009. *Design-Driven Innovation: Changing the Rules of Competition by Radically Innovating what Things Mean.* Boston: Harvard Business Review Press.

Vezzoli, Carlo and E. Manzini. 2008. *Design for Environmental Sustainability.* London: Springer.

Vitsoe. 2022. "Ethos – about Vitsoe." www.vitsoe.com, https://www.vitsoe.com/eu.

WBCSD. 2021. Vision 2050: Time to Transform – how Business can Lead the Transformations the World Needs: World Business Council for Sustainable Development.

Zhou, Liangchuan and Surendra Gupta. 2019. "Marketing Research and Life Cycle Pricing Strategies for New and Remanufactured Products." *Journal of Remanufacturing* 9 (1) (Apr 15): 29–50. doi:10.1007/s13243-018-0054-x. http://www.econis.eu/PPNSET?PPN=1668680084.

Index

Note: bold page numbers indicate tables; italic page numbers indicate figures.

renewing core principles strategy; and searching for hooks strategy 161, 165, **165**, 166–167, **166**, 167–168, **168**, 240–241, 244–245; *see also under specific products/companies*

differentiation 53–54, **107**, **108**, 232–233

dis-/reassembly, design for **10**, **12**, **15**

Dorst, K. 213, 227

durability 23, 35–36, 39; design for **10**, **12**, **15**, 104, **105**, 107; emotional *see* emotional durability; strategic *see* strategic durability

eco-design 16, 29

electronics, consumer 134–135

Ellen MacArthur Foundation 28

emotional durability 23, 33, 34, 36, 41, 107; design for 4, 11, **12**, **15**; and obsolescence 40–42

emotions 4, 11, 147, 150–151, 155

encourage sufficiency model 45, 46–47

energy use 32, 37

expert strategies 5–6, 88–94, 172–176; and action guides *see* action guides; application of 175–176, 188; business models (BMs) 43–47, 57–58; insights for practitioners 181–183; *see also specific strategies*

extended use, design strategies for **12**, 13, **15**

extending product value strategy 24, 90, 144–155; action guide for 155, 226–239; Coloplast SenSura Mio example 150–155; and competitive advantage 155, 227; and design principles 148–149, 151–154, **151**, **152**, **153**, **154**, 228–230, *228*; and differentiation 232–233; and emotions of users 146, 150–151, 155; and inspirational products 148–149, 153; and interview questions 238–239; and product-company fit 149, 236; and product-market fit 146, 147, *147*, 148–149, 226, 234–235; and product-user fit 149, 154, 226–227, 237; step 1 (identify ways to extend value) 147, 148, 226–227; step 2 (reframe fit with market) 147, 148–149, 227; step 3 (realign current strategic fit) 147, 149, 228–230

Faber, N.R. 28

flexibility, design for **12**, **15**

foreseeing future mismatches strategy 24, 90, 128–143; action guide for 143, 212–225; B&O A9 example 134–143; and design principles 137, 139, 140, 214–216, *214*, 224–225; and identifying mismatches 212, 218–219; and inspirational products 132, 133; and product-company fit 130, 133, 141–142, 143, 224; and product-market fit 133, 143, 225;

and product-user fit 130, *131*, 132–133, 137–141, 143, 212, 220–221; and stakeholder interviews 222–223; step 1 (identify changes in users' lifestyles) 131, 132, 134–137, *136*, 143, 212; step 2 (reframe strategic fit with users) 131, 132–133, 213; step 3 (renew strategic fit with company/market) 131, 133, 213–216

framing strategic fit *see* strategic fit, framing

ICT (information and communication technology) 37
innovation 55
inspirational products 117, 127, 132, 133, 148–149, 153, 161, 162, 240

Koopmans, Rudolf 29
Korhonen, J. 172
Kotler, Philip 54

LCA (lifecycle assessment) 33
LEGO Belville 72–73, 120–122, **123**, **124**
LEGO Friends series *7*, **19**, 22; and character development 122, 124–125; and company values 123–125; and core competencies 125; and 'friendship - all are equal' principle **123**, 125; and 'layers of colour/detail' principle 126, **127**; and leveraging objections strategy 24, 120–126, *121*; and 'pride of creation' principle 121, **122**; and product-company fit 72–74, *72*, *74*, 114, **122**, **123**, **124**, 127; and product-market fit 117, 121–122, **126**, 127; and product-user fit 117; and 'reuse across product lines' principle 120, **124**
LEGO Ninjago series *7*, **19**, 22, 84–86, *84*, *86*; and 'open and attractive story' principle 165, **165**; and 'power theme' principle 167–168, **168**; and product-company fit 158, 159, *159*, 160, 162, 167–168, **168**, 169; and product-market fit 158, 159, *159*, 160, 161–162, 166–167, **166**, 169; and product-user fit 158, 159, *159*, 160, 161, 164–166, **165**, 169; and searching for hooks strategy 24, 164–169; and starting from scratch 84–85, 159, 160, 169; and 'story with a tweak' principle 166–167, **166**
LEGO Scala 72, 73, 120–122, **123**, **124**
leveraging objections strategy 24, 90, 112–127, *115*; action guide for 127, 202–211; and company credibility 116, 121, **123**, 203–204; and company values 116, 123–125; and competitive advantage 116, 117; and core competencies 125; and core values/purpose/culture 116, 123; and design principles 117, 121–122, 127; and inspirational products 117, 127; LEGO Friends example 120–126, *121*; and product-company fit 114, **122**, **123**, **124**, 127, 202, 208; and product-market

fit 117, 121–122, **126**, 127; and product-user fit 117; and reuse across product platforms 120, **124**, 125; and stakeholder interviews 206–207; step 1 (identify objections/mismatches) 115, 116, 120–122, 202–204; step 2 (reframe mismatches) 115, 117, 122–125, 204–205; step 3 (test/align new strategic fit) 115, 117–118, 125–126, 205

lifecycle assessment (LCA) 33

linear economy 11, 50, 51, 57, 172, 175, 176

long use, design strategies for **12**, **15**

long-lasting strategic fit 23, 48–87, 172, 173, 181, 182, 184, 188; challenges with *see* strategic fit challenges; and company credibility 54, 56, *59*; and competitive advantage 50, 53–54, 56, 58, *59*, 102, 107; and conceptual design 50, 52–53, *52*; and core competencies/strategic strengths 55, 58, *59*; and core purpose/values 55–56, *59*; and framing *see* strategic fit, framing; and linear economy 51, *51*; map *59*; new, creation of *see* strategic fit, framing; and new product development 51–53, 62; and product-company fit *see* product-company fit; and product-market fit *see* product-market fit; and product-user fit *see* product-user fit; reframing *see* reframing strategic fit; and renewing strategic fit 90, 91, *91*; and strategic durability 5–6, 8, 9, 14, 57–58; strengths/weaknesses of 62, 76, 87, 90; transferring, challenge of 64, *64*, 67–70

long-term competitive fit 3–4, *44*

'magical interactions' principle **139**, 140–141, 173

maintenance/repair, design for 4, **10**, **12**, 13, **15**, *31*, 46

market conditions 8, *51*

memories/nostalgia 11

new markets, entering 51, 52, 83, *see also* searching for hooks strategy

niche markets 13, 18

obsolescence 23, 33, 37; and emotional durability 40–42; four types of 40–41, **40**; planned 41; resisting/postponing/reversing 11, **12**, 13, **15**, 35, 45, 50, 57, 172, 175

ostomy bags *see* Coloplast SenSura Mio

planned obsolescence 41

pollution 16, 28, 32

Porsche 45–46

Porras, Jerry I. 55

and renewing core principles strategy 100, 101, 103, **106**, 197; and searching for hooks strategy 158, 159, *159*, 160, 161, 162, 164–166, **165**, 169, 240–241, 242; and strategic fit challenge 191, *191*

167–168, **168**, 240–241, 244–245; and inspirational products 161, 162, 240; LEGO Ninjago example 164–169; and product-company fit 158, 159, *159*, 160, 162, 167–168, **168**; and product-market fit 158, 159, *159*, 160, 161–162, 166–167, **166**; and product-user fit 158, 159, *159*, 160, 161, 164–166, **165**; and starting from scratch 159, 160, 169; step 1 (framing product-user fit) 161, 164–166, 240–241; step 2 (framing product-market fit) 161–162, 166–167, 241; step 3 (framing product-company fit) 162, 167–168, 242

SenSura Mio *see* Coloplast SenSura Mio

stakeholders 43, 125–126, 148, 205, 226; interviews with 194, 200–201, 206–207, 215, 222–223, 229, 238–239

standardisation, design for **10**, **12**

starting from scratch *see* new markets, entering

Strappers, Pieter Jan 53

strategic conceptual design *see* conceptual design

strategic durability 4, 13–14, **15**, 16–17, 18, 23, 35–36, 90, 181, 184; and conceptual design 50; indicators **19**; and long-lasting strategic fit 5–6, 8, 9, 14, 17, 57–58, 87, 173, 174

strategic fit *see* long-lasting strategic fit

strategic fit challenges 60–87; creating new fit *see* strategic fit, framing; identifying 190–191; and product-company fit 62, *63*, 64, 71–74, 114; and product-market fit 62, *63*, 65, 79–82; and product-user fit 62, *63*, 65, 75–78; and strengths/weaknesses 62, 76, 87, 90; transferring long-lasting fit *see* strategic fit, transferring

strategic fit, framing 65, 83–86, 90, 93, *93*; LEGO Ninjago case study 84–86, *84*, *86*

strategic fit, reframing *see* reframing strategic fit

strategic fit, renewing 90, 91, *91*

strategic fit, transferring 64, *64*, 67–70, 91, 98; strategy for *see* renewing core principles strategy

strategic fit, weak 24, 62, 63, 64, 87, 92, 115, 116

strategic mismatches: future, foreseeing *see* foreseeing future mismatches strategy; identifying 115, 116, 120–122, 202–204, 218–219; product-company 62, 63, *63*, 71, 92, 114, 115, *115*, 117, 209; product-market 62, 63, *63*, 92, 121–122, 146, 147, *147*; product-user 62, 63, *63*, 75, 77, 92, *92*; reframing 115, 117, 122–125, 213

strategic strengths 55, 58, *59*, 90, 203

sustainability 2, 41, 50, 185; radical/relative approaches to 16, 28–29; and value 43

take-make-dispose pattern 11, 28, 33
technical possibilities 8
trust, design for 11, **12**, **15**
Tukker, Arnold 44

Ulwick, Anthony 53
upgradability, design for **10**, **12**, 13, **15**
user research 189, 195, 226–227

Valkenburg, R. 213, 227
value, product 24, *33*, 43; extending *see* extending product value strategy
value proposition (VP) 43–44, *44*
values, company *see* core values/purpose/culture
Van Doorsselaer, Karine 29
variability, design for 4, **10**
viable business 34
Vipp pedal bin 68–69, *103*, **105**, 106, 109, 110; and core principles 102–104; and product-market fit 102; and product-user fit 103, **106**
Vipp V1 kitchen *7*, **19**, 22; 'black trench coat' 102, 104, **105**; and competitive advantage 102, 107; and differentiation 107, 108; and durable design 104, **105**, 107; and 'Ford T-type choices' **108**, 109; and identifying core principles 102–104; and 'it's a tool' principle 103, 104–106, **106**; and 'nomad' kitchen concept 107–109, **107**; and product-company fit 100, 101; and product-market fit 102, **107**, **108**; and product-user fit 100, 101, 103, **106**; and renewing core principles strategy 23, 102–110, *103*; and renewing/implementing core principles 104–106; and strategic fit 68–70, *68*, *70*; and strengthening strategic fit 106–109
Vitsoe (furniture company) 46–47

waste 41; minimisation 11, 16